MAJOR REVISION FACTS IN MATHEMATICS

B. N. Kumar

T0127911

University Press of America,® Inc.
Lanham · Boulder · New York · Toronto · Plymouth, UK

**Copyright © 2009 by
University Press of America,® Inc.**
4501 Forbes Boulevard
Suite 200
Lanham, Maryland 20706
UPA Acquisitions Department (301) 459-3366

Estover Road
Plymouth PL6 7PY
United Kingdom

Library of Congress Control Number: 2009931882
ISBN: 978-0-7618-4784-7 (paperback : alk. paper)
eISBN: 978-0-7618-4785-4

Table of Contents

Purpose and Audience

Major Revision Facts form the basis for recognizing the key concepts and apply them for examination answers. It covers the basic areas of the Regents Mathematics, SAT 11, and open entrance College examinations, as well as first year College Mathematics. They are ideal for a quick grasp of the subject and for revision but should also be used as a reference throughout the course of study.

Major Revision Facts are compiled after detailed firsthand knowledge and experience of teaching Mathematics, classroom experience and firsthand knowledge of examining and examinations.

Major Revision Facts are suitable for students studying for examinations at school and college.

Major Revision Facts cover the most important facts of the subject syllabus for the various school and college examinations.

Major Revision Facts can be used for learning in spare moments. They can be used along with basic texts.

Major Revision Facts make study time more productive, reinforcing the major concept

After all, no one can pass an examination for you. Your success depends on how you apply yourself and how much effort you are able to make.

Therefore, LEARN the major facts thoroughly. Work through one section each day. Review regularly throughout your course is far better than last minute cramming.

Always think about the work. Try to comprehend and internalize it. Grasp the interrelationships and explanations. Be sure to know the concepts covered and always use your textbooks for fuller details if this seems necessary.

ALWAYS HAVE YOUR MAJOR REVIEW FACTS WITH YOU. LET THEM HELP YOU TO EXAM SUCCESS. USE THEM WISELY.

Introduction

Major Revision Facts in Mathematics has been tried out and used extensively with students both locally and overseas. They have been used in Math teacher workshops and seminars. To a very large extent they have been used at the Hillside Learning Center, Queens, New York. A cross-section of student population including those taking the Regents Board Examination, SAT II, and College Mathematics benefited extensively from the special programs.

Theory of Quadratics

Quadratic Equations

The general quadratic equation is $ax^2 + bx + c = 0$ and its roots (i.e. the values of x satisfying the equation) are given by:

$$x = \frac{-b \pm \sqrt{b^2 - 4ac}}{2a}$$

$b^2 - 4ac$ is called the **discriminant (Δ)** and determines the nature of the roots.

   I.   If $\Delta > 0$, the roots are **real** and **unequal**. If it is a **perfect square** the roots are **rational**; otherwise the roots are **irrational**.

   II. If $\Delta = 0$, the roots are **real, equal** and **rational.**

   III. If $\Delta < 0$, the roots are **unreal** (or imaginary).

Example: $x^2 - 1 = p(x - 1)$ can be rearranged as
$x^2 - px + (p - 1) = 0$ and has real roots if $\Delta \geq 0$.
$\Delta = (-p)^2 - 4.1(p - 1) = p^2 - 4p + 4 = (p - 2)^2 \geq 0$ **for real values of p.**
i.e. the roots of the equation are real if p is real.

Geometrically, **I** is the case when the graph of $y = ax^2 + bx + c$ cuts the x-axis in two points; **II** when the graph touches the x-axis; **III** when the graph does not cut the x-axis.

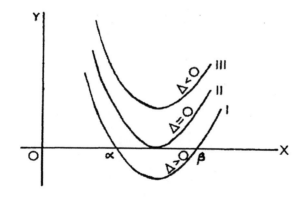

SUM AND PRODUCT OF THE ROOTS

α and β are the roots of $(x - \alpha)(x - \beta) = 0$
i.e. of $x^2 - (\alpha + \beta)x + \alpha\beta = 0$.
If they are also the roots of $ax^2 + bx + c = 0$
then $x^2 - (\alpha + \beta)x + \alpha\beta = 0$

and $x^2 + \frac{b}{a}x + \frac{c}{a} = 0$ are the same equation.

i.e. $\alpha + \beta = -\frac{b}{a}$ and $\alpha\beta = \frac{c}{a}$

To form other equations whose roots α', β' are functions of α and β, express S, the **sum** of α' and β', in terms of $\alpha + \beta$ and $\alpha\beta$ and substitute for $\alpha + \beta$ and $\alpha\beta$ in terms of the coefficients a, b, and c.

The required equation is
$$x^2 - Sx + P = 0$$
which must be cleared of fractions if necessary. Note that α and β are not separately found.

Two useful relations: $\alpha^2 + \beta^2 = (\alpha + \beta)^2 - 2\alpha\beta$
$\alpha^3 + \beta^3 = (\alpha + \beta)^3 - 3\alpha\beta(\alpha + \beta)$

Example: If α, β are the roots of $x^2 - x + 3 = 0$, find the equation whose roots are $2\alpha + \beta$, $\alpha + 2\beta$.

$a = 1$, $b = -1$, $c = 3$; $\alpha + \beta = -\frac{b}{a} = 1$, $\alpha\beta = \frac{c}{a} = 3$.
$S = (2\alpha + \beta) + (\alpha + 2\beta) = 3(\alpha + \beta) = 3$
$P = (2\alpha + \beta)(\alpha + 2\beta) = 2\alpha^2 + 5\alpha\beta + 2\beta^2$
$\quad = 2(\alpha + \beta)^2 + \alpha\beta = 2 + 3 = 5$
Required equation is $x^2 - 3x + 5 = 0$

QUADRATIC FUNCTIONS

The quadratic function $ax^2 + bx + c$ is equal to zero for two real values of x, α and β say, only if $\Delta \geq 0$ and in this case the function can be written $a(x - \alpha)(x - \beta)$. If x is **greater than the larger** of α and β both brackets are positive, and if x is **less than the smaller** of α and β both brackets are negative. In either case the product of the brackets is **positive** and the function has the same sign as a. If x has a value **between** α and β, one bracket is positive and the other is negative, and their product is therefore **negative** and so the function has a sign opposite to that of a.

Example: $2x^2 - 5x - 3 = (2x + 1)(x - 3)$

$\qquad = 2(x + \frac{1}{2})(x - 3) = 0$ when $x = -\frac{1}{2}$ or when $x = 3 > 0$ when $x > 3$

$\qquad$ Or when $x < -\frac{1}{2} < 0$ when $-\frac{1}{2} < x < 3$.

If $\Delta = 0$, the function is zero for only one real value of x, α say, and can be written $a(x - \alpha)^2$. $(x - \alpha)^2$ is **positive for all real values of x**(except $x = \alpha$) and the function has the same sign as a for all values of x except $x = \alpha$. In this case the value of the function is zero.

Example: $4x^2 - 12x + 9 = (2x - 3)^2 = 4(x - \frac{3}{2})^2$

$$= 0 \text{ when } x = \frac{3}{2}$$

$$> 0 \text{ for all other real values of } x.$$

If $\Delta < 0$, the function is **not** equal to zero for any real value of x and is **positive if "a" is positive** and is **negative if "a" is negative**.

Example: $x^2 - x + 1 > 0$ for all real values of x since $\Delta = (-1)^2 - 4 = -3 < 0$ and the coefficient of x^2 is positive.

Geometrically, the graph of the function is convex downward for positive values of a and convex upwards for negative values of a. The three possibilities are illustrated previously on page 1 for positive values of a.

MAXIMUM AND MINIMUM VALUES

The quadratic function $ax^2 + bx + c$ has a least value when a is positive and a greatest value when a is negative. The maximum or minimum value can be obtained by writing the function in the form $\mathbf{a(x + p)^2 + q}$ and then putting $\mathbf{x + p = 0}$.

Example: Find the minimum value of $2x^2 + 4x - 1$.

$$2x^2 + 4x - 1 = 2(x + p)^2 + q$$
$$= 2x^2 + 4px + 2p^2 + q$$

Hence $\qquad 4p = 4$ and $2p^2 + q = -1$

$\therefore \qquad p = 1$ and $q = -3$

Thus the minimum value of $2x^2 + 4x - 1$ is -3.

Note how the values of p and q are found by putting the coefficients of x equal and the constant terms equal.

Progressions and Series

A **progression** (or sequence) is a set of numbers following one another in some defined way, e.g. 1, 3, 5, ...

A sequence in which each term is derived from the previous term by **adding** (or subtracting) a **constant amount,** called the **common difference (d),** is an **Arithmetic Progression (A.P.).** Denoting the general term, or nth term, by T_n and putting $T_1 = a$, we have

$$T_2 = a + d, T_3 = a + 2d, \ldots T_n = a + (n - 1)d$$

Writing $T_n = l$, the **sum of the first n terms, S_n,** can be written in forward and reverse order as follows:

$$a + (a + d) + (a + 2d) + \ldots + (l - 2d) + (l - d) + l$$

$$l + (l - d) + (l - 2d) + \ldots + (a + 2d) + (a + d) + a$$

Adding and noting that each of the n pairs sums to $(a + l)$, and dividing by 2 we have:

$$S_n = \frac{n}{2}(a + l) = \frac{n}{2}\{2a + (n - 1)d\}$$

Example: Find the sum of all the numbers between 1 and 200 divisible by 5.

$$T_1 = a = 5, d = 5, T_n = a + (n - 1)d = 195$$

$$\therefore\ 5 + (n - 1).5 = 5n = 195 \text{ and } n = 39$$

$$\therefore\ S_{39} = \frac{39}{2}(5 + 195) = 39 \times 100 = 3900$$

If a, b, c are in A.P., b is the **Arithmetic Mean** of a and c. Since $d = b - a = c - b$, then $\mathbf{b = \frac{1}{2}(a + c)}$.

A sequence in which each term is obtained from the previous term by **multiplying** (or dividing) by a **constant ratio,** called the **common ratio (r),** is a **Geometric Progression (G.P.).** For a G.P. we have:

$$T_1 = a, T_2 = ar, T_3 = ar^2, \ldots T_n = ar^{n-1}$$

The sum of the first n terms, S_n, can be written:

$$S_n = a + ar + ar^2 + \ldots + ar^{n-2} + ar^{n-1}$$

$$r \times S_n = ar + ar^2 + \ldots + ar^{n-2} + ar^{n-1} + ar^n$$

Subtracting and dividing by $1 - r$, we obtain

$$S_n = \frac{a(1 - r^n)}{1 - r}$$

Example: Find the sum of the first six terms of the G.P. 3, 6, 12,…

$a = 3, r = 2$, hence $S_6 = \dfrac{3(1-2^6)}{1-2} = \dfrac{3(1-64)}{-1} = 189$

If **-1< r < 1**, r^n tends to 0 as n tends to infinity, S_n tends to S_∞, the **sum to infinity**, and $S_\infty = \dfrac{a}{1-r}$

Example: The G.P. $1 - \dfrac{1}{2} + \dfrac{1}{4} - \ldots$ has a sum to infinity because $r = -\dfrac{1}{2}$ which lies between -1 and 1 and

$$S_\infty = \frac{1}{1-(-\frac{1}{2})} = 1\,\frac{1}{2}\Big/\frac{3}{2} = \frac{2}{3}$$

If a, b, c are in G.P., b is the **Geometric Mean** of a and c.
Since $ar^2 = br = c$,
Then $b^2 = ac$, or $\mathbf{b} = \pm\sqrt{ac}$

A **series** is formed by adding the terms of a sequence in order. $1 + 2 + 3 + 4+\ldots$ is an arithmetic series since the terms are in A.P. and $1 + 2 + 4 +8 +\ldots$ is a geometric series since the terms are in G.P.

THE BINOMIAL SERIES

If n is a positive integer the **Binomial Theorem** states

$$(\mathbf{a + x})^n = \mathbf{a}^n + \frac{n}{1}.\,\mathbf{a}^{n-1}.\,\mathbf{x} + \frac{n(n-1)}{1.\,2}.\,\mathbf{a}^{n-2}.\,\mathbf{x}^2 + \frac{n(n-1)(n-2)}{1.\,2.\,3}.\,\mathbf{a}^{n-3}.\,\mathbf{x}^3 + \ldots + \mathbf{x}^n$$

This is a **terminating** series having $(n + 1)$ terms. The seconds term, i.e. x, may be negative or may be replaced by a more complex term or group of terms.

Example: Expand $(1 + 2x)^5$
Here $a = 1$ and x is replaced by $2x$. Hence:

$$(1 + 2x)^5 = 1^5 + \frac{5}{1}.\,1^4.\,(2x) + \frac{5\times4}{1\times2}.\,1^3.\,(2x)^2 + \frac{5\times4\times3}{1\times2\times3}.\,1^2.\,(2x)^3 + $$

$$\frac{5\times4\times3\times2}{1\times2\times3\times4}.\,1.\,(2x)^4 + (2x)^5 =$$

$$1 + 10x + 40x^2 + 80x^3 + 80x^4 + 32x^5$$

A particular term is best found by writing out the expansion until the required term is reached.

Example: Find the term independent of x in $(x + x^{-1})^6$.
Here a and x are replaced by x and x^{-1} respectively.

$$(x + \tfrac{1}{x})^6 = x^6 + \tfrac{6}{1} \cdot x^5 \cdot \tfrac{1}{x} + \tfrac{6 \times 5}{1 \times 2} \cdot x^4 \cdot (\tfrac{1}{x})^2 + \tfrac{6 \times 5 \times 4}{1 \times 2 \times 3} \cdot x^3 \cdot (\tfrac{1}{x})^3 +$$

Required term is the fourth, i.e. $\dfrac{6 \times 5 \times 4}{1 \times 2 \times 3} \cdot x^3 \cdot \dfrac{1}{x^3} = 20$.

When n is a **negative integer or a positive or negative fraction,** the expansion still holds but with two important restrictions:

    (1) **a must equal , (2) x must lie between − 1 and +1.** In these cases the series does not terminate, i.e. it is an infinite series. The series is then:

$$\mathbf{(1 + x)^n = 1 + \tfrac{n}{1} \cdot x + \tfrac{n(n-1)}{1 \times 2} \cdot x^2 + \tfrac{n(n-1)(n-2)}{1 \times 2 \times 3} \cdot x^3 + ...}$$

Example: If $-1 < x < 1$, $\dfrac{1}{1+x} = (1 + x)^{-1}$

$$= 1 + \tfrac{-1}{1} \cdot x + \tfrac{(-1)(-1-1)}{1 \times 2} \cdot x^2 + \tfrac{(-1)(-1-1)(-1-2)}{1 \times 2 \times 3} \cdot x^3 + ...$$

$$= 1 - x + \tfrac{(-1)(-2)}{1 \times 2} \cdot x^2 + \tfrac{(-1)(-2)(-3)}{1 \times 2 \times 3} x^3 + ...$$

$$= 1 - x + x^2 - x^3 + ...$$

APPROXIMATIONS

When **x is small** the **higher powers of x are very small** and so the sum of the first few terms of the Binomial Series gives a good approximation in such cases.

Example: Evaluate 1.01^6 correct to four places of decimals.

$$(1 + x)^6 = 1 + 6x + 15x^2 + 20x^3 + 15x^4 + 6x^5 + x^6$$

Substituting $x = 0.01$ and working to two more figures than required in the final answer and neglecting terms too small to affect the desired approximation,

$$1.01^6 = 1 + 0.06 + 0.0015 + 0.000020 = 1.061520$$
$$= 1.0615 \text{ correct to four places of decimals}$$

Simultaneous Equations

The solution of simultaneous linear equations in **three unknowns** requires **three** equations. The procedure is:

(1)   Take the equations in **two pairs** and **eliminate one of the unknowns** to get two equations in two unknowns.

(2)   Solve these two equations in the usual way.

(3)   Find the remaining unknown by **substitution** in one of the **Original** equations.

(4)   **Check** the solutions in the other **two** equations.

Example: Solve

$$x + y + z = 2 \tag{1}$$
$$2x - y + z = 7 \tag{2}$$
$$3x - 4y - 2z = 5 \tag{3}$$

(1) + (2), to eliminate y: $\quad 3x + 2z = 9 \tag{4}$

(1) x 4 + (3), to eliminate y: $\quad 7x + 2z = 13 \tag{5}$

(5) − (4), to eliminate z: $\quad 4x = 4, \therefore x = 1$

Substitute for x in (4): $\quad 2z = 9 - 3, \therefore z = 3$

Check in (5).

Substitute for x and z in (1) to obtain $y = -2$

Check in (2) and (3).

For two equations in two unknowns when one equation is **linear** and the other **quadratic** the method is:

(1)   Use the **linear** equation to express one unknown in terms of the other.

(2)   Substitute in the **quadratic** equation and solve.

(3)   Find the other unknown by **substitution** in the **linear** equation.

(4)   **Check** the **pairs** of solutions in the **quadratic** equation.

Example: Solve

$$x - 2y = 1 \tag{1}$$
$$x^2 + 2xy + y^2 = 16 \tag{2}$$

From (1): $\quad x = 2y + 1$

Substitute for x in (2):

$$(2y + 1)^2 + 2(2y + 1)y + y^2 = 16$$

Expanding, collecting terms, and simplifying gives

$$3y^2 + 2y - 5 = 0$$

$\therefore (y - 1)(3y + 5) = 0$ and hence $y = 1$ or $= 1$ or $y = -1\frac{2}{3}$

Substitute for y in (1) to obtain $x = 3$ when $y = 1$ and $x = -2\frac{1}{3}$ when $y = -1\frac{2}{3}$.

$x = -2\frac{1}{3}$ when $y = -1\frac{2}{3}$.

Check each pair of solutions in (2).

The lines that have been omitted from these outline solutions should be inserted by the student.

Indices and Logarithms

When x and y are **positive integers** it is easily proved by writing out the terms in full that:

(I) $\quad a^x \times a^y = a^{x+y}$

(II) $\quad a^x/a^Y = a^{x-Y} \ (x > y)$

(III) $\quad (a^x)^y = a^{xy}$

Since $a^x / a^x = a^{x-x} = a^0$ and $a^x / a^x = 1$, hence

(IV) $\quad a^0 = 1.$

By requiring (I) − (IV) to hold for all x and y we obtain meanings for negative and fractional indices.

Since $a^{-2} \times a^2 = a^{-2+2} = a^0 = 1$, $a^{-2} = 1/a^2.$

In general, $\mathbf{a^{-n} = 1/a^n = }$ **the reciprocal of $\mathbf{a^n}$.**

Since $a^{1/2} \times a^{1/2} = a^{1/2 + 1/2} = a^1 = a = \sqrt{a} \times \sqrt{a}$, $a^{1/2} = \sqrt{a}$

In general, $a^{1/n} = \sqrt[n]{a} \qquad = $ **nth root of a, and $\mathbf{a^{m/n} = (a^{1/n})^m}$**

$$= (\sqrt[n]{a})^n = \textbf{nth root of } a \textbf{ to the power } m$$

If $M = a^x$ we define x as the **logarithm of M to the base a** and write $x = \log_a M$. If $N = a^Y$ then $y = log_a N$ and corresponding to (I)-(IV) we have:

(I) $\quad M \times N = a^x \times a^y = a^{x+y}$

$\quad \therefore x + y = \log_a (M \times N) = \log_a M + \log_a N.$

(II) $\quad M/N = a^x/a^y = a^{x-y}$

$\quad \therefore x - y = \log_a(M/N) = \log_a M - \log_a N.$

(III) $\quad M^P = (a^x)^P = a^{xP}, \quad \therefore \ px = \log_a(M^P) = p \times \log_a M.$

(IV) $\quad a^0 = 1, \quad \therefore \ \log_a 1 = 0$ **whatever the base.**

In addition, since $a^1 = a \ \therefore \ \log_a a = 1$ for all a.

To find the logarithm of M to another base b, let $z = \log_b M$. Then $M = b^z$ and $\log_a M = \log_a b^z = z \times \log_a b.$

Hence the **base changing rule, $z = \log_b M = \dfrac{\log_a M}{\log_a b}$**

Equations such as $2^x = 5$ are solved by taking logs.

Thus $\log_{10} 2^x = x . \ \log_{10} 2 = \log_{10} 5$

So, $x = 0.6990/0.3010 = 2.32$

Remainder Theorem

The Remainder Theorem states that if **any polynomial P(x) is divided by (x —k)** until the remainder R contains no term in x, **R is equal to the value of P(x) when x = k, i.e. P(k).**
If $Q(x)$ is the quotient on division by $(x - k)$,

$$P(x) \equiv (x - k) \times Q(x) + R$$

Substituting $x = k$: $P(k) = O. Q(k) + R$, i.e. $R=P(k)$.
Thus when $P(x) \equiv 2x^3 + x^2 - 3x$ is divided by $(x + 1)$ the remainder $= P(-1) = 2.(-1)^3 + (-1)^2 - 3.(-1) = 2.$
Many examples in which the Reminder Theorem is used lead to simultaneous equations.

Example: When $ax^3 - x^2 + bx - 1$ is divided by $(x - 1)$ the remainder is 3; when divided by $(x - 2)$ the remainder is 23. Find the values of a and b.

Let $P(x) \equiv ax^3 - x^2 + bx - 1.$
By the Remainder Theorem:

$P(1)= a-1+b-1 = 3,$	i.e. $a+b = 5$
$P(2) = 8a-4+2b-1 = 23,$	i.e. $8a + 2b = 28$
Eliminating b: $3a = 9,$	i.e. $a = 3.$
Substituting for a gives	$b = 2.$

FACTOR THEOREM

If P(k) = 0, there is no remainder and it follows that **(x —k) is a factor of P(x).**
This is the Factor Theorem; it is particularly useful in the **solution of equations of the third or higher degree** when some of the roots are **simple integers or rational numbers.**
Example: Solve $x^3 + 2x^2 - 2x - 3 = 0.$

Let $P(x) \equiv x^3 + 2x^2 - 2x - 3.$
The only values of x which can make $P(x) = 0$ are the **factors of the number term,** i.e. $\pm 1, \pm 3.$
By the Factor Theorem:
$P(1) = 1 + 2 - 2 - 3 \neq 0$, i.e. $x - 1$ is not a factor.
$P(-1) = -1 + 2 + 2 - 3 = 0$, i.e. $x + 1$ is a factor.
Hence $P(x) \equiv (x + 1) (x^2 + x - 3)$, the second factor being obtained by dividing $P(x)$ by $(x + 1).$
$\therefore P(x) = 0$ when $x + 1 = 0$ or when $x^2 + x - 3 = 0.$
Hence $x = -1, 1.30, -2.30.$ (Last 2 roots by formula.)

Permutations and Combinations

When a **choice** is made from a set of entities it is essential to distinguish between the case in which the choice involves an **arrangement** of the chosen entities and that in which a **group is selected without regard to the order of selection**. The choice involving an arrangement is a **permutation**. The selection of a group in which the elements are not to be arranged among themselves is a **combination**. In either case the choice may be of **all or only some** members of the set. Problems on permutations and combinations are better dealt with in the ways illustrated in the following examples than by formulae.

Example 1: The number of numbers that can be made from the digits 1, 2, 3, 4 if no digit can be used twice. The first digit can be any one of 1, 2, 3, 4 and so can be chosen in 4 ways:

i.e. there are *4* 1-digit numbers.

The second digit can be chosen from the 3 digits left in 3 ways; each first - choice can be taken with each second choice so that the two places are filled in 4.3 ways;

i.e. there are 4 x 3 = 12 2-digit numbers.

The third digit can then be chosen in 2 ways:

i.e. there are 4 x 3 x 2 = 24 3-digit numbers.

The fourth digit can now be chosen in only one way:

i.e. there are 4 x 3 x 2 x 1 = 24 4-digit numbers.

These steps show the number of permutations that can be made from 4 unlike elements when taken 1 at a time, 2 at a time, 3 at a time and 4 at a time. The number of numbers formed = 4 +12 +24 +24 = 64.

The **product of successive integers starting with 1** is given a symbol (!) and a name **(factorial)**.

Thus 1 x 2 x 3 x 4 = 4! and reads 4 factorial.

Example 2: In how many arrangements using all the letters of the word 'pound' are the vowels separated?

The 5 letters can be arranged in 5! = 5 x 4 x 3 x 2 x 1 = 120 ways. Treating the vowels as one letter, the number of arrangements = 4! = 24, in each of which the vowels can now be arranged in 2 ways as ou or uo.

∴ the number of arrangements with the vowels together
= 2 x 24 = 48.

∴ the number of arrangements with the vowels separated
= 120 — 48 = 72.

Example 3: In how many ways can a committee of seven be seated at a round table?

This is a **circular** permutation in which the first choice is a **random** one and the subsequent choices are made in relation to the first. The first member having sat down at random, the second member has 6 choices, the third has 5 choices, and so on. The number of seating arrangements is, therefore, $6! = 6 \times 5 \times 4 \times 3 \times 2 \times 1 = 720$.

Example 4: Find the number of arrangements of the letters of the word PARALLEL.
This is a permutation complicated by **not having all the elements distinct**. If all the eight letters were different the number of permutations would be $8!$.
Considering any one permutation, the 2 A's can be interchanged in $2!$ ways and the 3 L.'s in $3!$ ways without altering the permutation. Hence each permutation appears $2! \times 3!$ times in the $8!$ ways.

$\therefore$ The number of distinct permutations $= \dfrac{8!}{2! \times 3!}$

Example 5: How many triangles can be formed by joining the vertices of an octagon?
Here we are concerned with a **combination** for we have to select a group of 3 points from 8 points. The 3 points can be selected from the 8 in $8 \times 7 \times 6$ ways which include all possible orders of choice. Hence each set of 3 points appears $3!$ times in the $8 \times 7 \times 6$ ways.

$\therefore$ **Number of triangles formed** $= \dfrac{8 \times 7 \times 6}{3!}$

Example 6: How many different forecasts are possible for the results of six football matches each of which may end in a win, a loss or a draw?
The result of any match is **independent** of the result of any other. Each match can end in 3 possible ways which can be combined with any one of the 3 possible results for every other match.

$\therefore$ Number of different forecasts $= 3^6$

The general formula for the number of permutations of n distinct objects taken p at a time is
$$_nP_p = \frac{n!}{(n-p)!}$$
The number of combinations of n objects taken p at a time is

$$_nC_p = \frac{n!}{p!(n-p)!}$$

Graphs of Algebraic Functions

The object of **sketching** a curve is to show its **general shape** and **special features**. The curve is **not** drawn strictly to scale. A good idea of the shape is gained by the following procedure, in full or in part.

I.   Find where the curve **crosses** the axes, i.e. consider $x = 0$ and $y = 0$.

II.   Decide if the curve is **symmetrical** about either axis. If the equation contains only even powers of x the curve is symmetrical about the y-axis; if it contains only even powers of y the curve is symmetrical about the **X**-axis.

III.   Determine if there are values of either variable for which there are **no real values** of the other.

IV.   Consider the value of y for **large positive** and **large negative** values of x.

V.   Consider the value of y for **small positive** and **small negative** values of x.

VI.   Find the **turning points**.

Example: Sketch the curve $y = (x + 1) (3 - x)$.

I.   When $x = 0, y = 3$. When $y = 0, x = -1$ or 3.

II. $y = 3 + 2x - x^2$ so the curve is not symmetrical about either axis. The equation can be written $4 - y = (1 - x)^2$ so the curve is symmetrical about the line $x = 1$.

III. y is real for all real values of x. Since $(1 - x)^2$ is always positive or zero, $4 - y \geq 0$ and so $y \leq 4$. There are no real values of x for values of y greater than 4.

IV.   When x is large the value of y depends mainly on $-x^2$ and so y is large and negative for both large positive and large negative values of x.

V.   Does not apply in this example

VI. $\frac{dy}{dx} = 2 - 2x = 0$ when $x = 1$, and $\frac{d^2 y}{dx^2} = -2 < 0$.

∴ (1,4) is a maximum turning point.

It was not really necessary to use VI as I-IV gives enough information to sketch the graph. (*Figure 1 on next page*).

A sketch of a curve can help to avoid making incorrect statements or offer a check on errors in calculation, e.g. in a confusion of maximum and minimum points. The student should be familiar with curves like that of the example and the curves of which $y = x^3$, $xy = 1$ and $y = x^3 - 2x^2 - 5x + 6$ are typical equations.

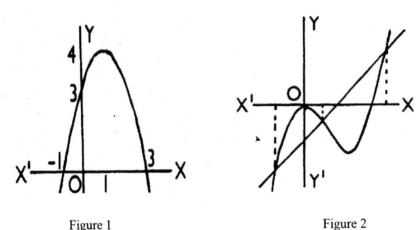

Figure 1   Figure 2

GRAPHICAL SOLUTIONS

Accurately drawn graphs are used to obtain **approximate solutions** of equations. To find the roots of the equation $x^3 - 2x^2 - x + 1 = 0$ we can draw the graph of $y = x^3 - 2x^2 - x + 1$ and find the **points of intersection with the x-axis.** Alternatively, we can draw the graphs of $y = x^3 - 2x^2$ and $y = x - 1$ using the same scale and axes and read off the **value of x at the intersections of the graphs.** At these points both graphs have the same value of y so $x^3 - 2x^2 = x - 1$, i.e. $x^3 - 2x^2 - x + 1 = 0$ at these points. (*Fig. 2*).

The following points should be observed carefully:

I.   If the scale is not given choose a scale to give the **largest** graph possible and **units easily sub-divided** by the lines of the graph paper.

II.   Find and tabulate the values of y corresponding to values of x to obtain a **sufficient** number of points, particularly where the curve is **bending sharply** and where v is **changing rapidly.**

III.   Show how answers are obtained and read the answers off as **accurately** as possible.

Circular Measure

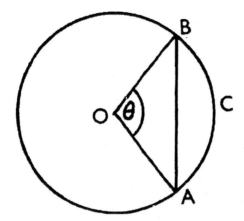

A radian is the **angle** subtended at the **centre** of a circle by an **arc of length equal to the radius.**

Circumference of a circle $= 2\pi$ times the radius.

∴ the angle subtended at the centre by the circumference
$= 2\pi$ radians $= 360°$.

∴ π **radians** $= 180°$. Hence $\theta° = \dfrac{\theta\pi}{180}$ radians.

Thus $90° = \dfrac{\pi}{2}$ radians, $60° = \dfrac{\pi}{3}$ radians, etc.

If an arc AB of a length s subtends an angle θ radians at the center 0 of a circle or radius r.

$$\frac{\text{length of arc AB}}{\text{circumference}} = \frac{s}{2\pi r} = \frac{\theta}{2\pi}, \dots s = r\theta$$

Also, $\dfrac{\text{area of sector AOB}}{\text{area of circle}} = \dfrac{A}{\pi r^2} = \dfrac{\pi}{2\pi}$

∴ **area of sector** $= \dfrac{1}{2}r^2\theta$

Area of segment ACB $=$ area sector AOB $-$ area $\triangle AOB$

$$= \frac{1}{2}r^2\theta - \frac{1}{2}r^2\sin\theta = \frac{1}{2}r^2(\theta - \sin\theta).$$

Since 0 is in radians, the radians must be converted into degrees before using the trigonometric tables.

Example: Find $\sin x$ when $x = 0.5$ radians.

$$0.5 \text{ radians} = \frac{0.5 \times 180^{\circ}}{\pi} = \frac{0.5 \times 180^{\circ}}{3.142} = 28^{\circ}39'.$$

$$\therefore \sin x = \sin 28^{\circ}39' = 0.4795.$$

The standard formulae for derivatives and integrals of trigonometric functions are given for the **angle in radians.**

Trigonometric Functions

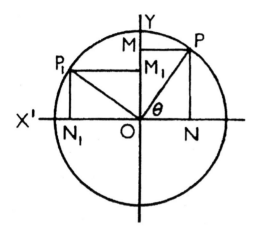

$$\mathbf{Y^1}$$

An angle θ of any magnitude can be represented by the angle between a radius OP of a circle, centre the origin and radius 1, and the positive OX direction. Positive angles are measured counter-clockwise from OX and negative angles are measured clockwise from OX.

To be consistent with the definitions of sine, cosine and tangent for acute angles, we define:

$sin\theta$ = projection of OP on OY = y coordinate of P
$cos\theta$ = projection of OP on OX = x coordinate of P

$$tan\theta = \frac{\text{projection of OP on OY}}{\text{projection of OP on OX}} = \frac{sin\,\theta}{cos\,\theta}$$

The **projections on the axes have the usual sign conventions.** The following relations are easily deduced from the definitons:

	$90° < \theta° < 180°$	$180° < \theta° < 270°$	$270° < \theta° < 360°$
$sin\,\theta$	$+sin\,(180° - \theta)$	$-sin\,(\theta - 180°)$	$-sin\,(360° - \theta)$
$cos\,\theta$	$-cos\,(180° - \theta)$	$-cos\,(\theta - 180°)$	$+cos\,(360° - \theta)$
$tan\,\theta$	$-tan\,(180° - \theta)$	$+tan\,(\theta - 180°)$	$-tan\,(360° - \theta)$

$\sin(-\theta) = -\sin\theta$, $\cos(-\theta) = \cos\theta$, $\tan(-\theta) = -\tan\theta$.

$\sin 90° = 1$, $\cos 90° = 0$, $\tan 90° = \infty$;

$\sin 180° = 0$, $\cos 180° = -1$; $\tan 180° = 0$;

$\sin 270° = -1$, $\cos 270° = 0$, $\tan 270° = \infty$.

For, angles greater than 360° first subtract the multiples of 360°, e.g. $\sin 930° = \sin 210° = -\sin 30°$.

The remaining trigonometric functions are defined as:

$$\mathbf{cosec\theta = \frac{1}{sin\theta}, \ sec\theta = \frac{1}{cos\theta}, \ cot\theta = \frac{1}{tan\theta} = tan(90° - \theta)}$$

Addition Formulae

The following formulae are true for all A and B.

$\sin (A + B) = \sin A \times \cos B + \cos A \times \sin B$

$\sin (A - B) = \sin A \times \cos B - \cos A \times \sin B$

$\cos (A + B) = \cos A \times \cos B - \sin A \times \sin B$

$\cos (A - B) = \cos A \times \cos B + \sin A \times \sin B$

Example. $\sin 75° = \sin (45° + 30°)$

$$= \sin 45°. \cos 30° + \cos 45° . \sin 30°$$

$$= \frac{\sqrt{2}}{2} \times \frac{\sqrt{3}}{2} + \frac{\sqrt{2}}{2} \times \frac{1}{2} = \frac{1}{4}(\sqrt{6} + \sqrt{2}).$$

$$\tan(A + B) = \frac{\sin(A+B)}{\cos(A+B)} = \frac{\sin A \times \cos B + \cos A \times \sin B}{\cos A \times \cos B - \sin A \times \sin B}$$

Dividing every term by cosA x cosB we obtain:

$$\textbf{Tan}(A + B) = \frac{\tan A + \tan B}{1 - \tan A \times \tan B}$$

Similarly $\qquad \textbf{Tan}(A - B) = \frac{\tan A - \tan B}{1 + \tan A \times \tan B}$

Writing B = A in the above formula we obtain:

$\text{Sin} 2A = a \sin A \times \cos A$

$\text{Cos} 2A = \cos^2 A - \sin^2 A = 2\cos^2 A - 1 = 1 - 2\sin^2 A$

$\qquad$ (on using $\cos^2 A + \sin^2 A = 1$)

$$\text{Tan} 2A = \frac{2 \tan A}{1 - \tan^2 A}$$

Example: $\sin 3\theta = \sin (2\theta + \theta) = \sin 2\theta \times \cos \theta + \cos 2\theta \times \sin \theta = $
$2 \sin \theta \times \cos^2 \theta + (1 - 2 \sin^2 \theta) \times \sin \theta$
$= 2 \sin \theta \times (1 - \sin^2 \theta) + \sin \theta - 2 \sin^3 \theta = 3 \sin \theta - 4 \sin^3 \theta.$
$\quad$ **Adding the expansions** for sin $(A + B)$ and sin $(A - B)$ we have sin
$(A + B) + \sin (A - B) = 2 \sin A x. \cos B$

Put $S = A + B, D = A - B$, so $A = (\frac{1}{2}S + D), B = \frac{1}{2}(S - D)$

Then $\qquad$ $\sin S + \sin D = 2 \sin\frac{1}{2}(S + D) \times \cos \frac{1}{2}(S - D)$

Similarly $\quad$ $\sin S - \sin D = 2 \sin\frac{1}{2} (S - D) \times \cos \frac{1}{2} (S + D)$

$$\cos S + \cos D = 2 \cos \frac{1}{2} (S + D) \times \cos\frac{1}{2} (S - D)$$

$$\cos S - \cos D = -2 \sin \frac{1}{2} (S - D) \times \sin\frac{1}{2} (S + D)$$

Example: $\dfrac{sin50^{\circ}-sin10^{\circ}}{cos10^{\circ}-cos50^{\circ}}$

$$= \dfrac{2sin\frac{1}{2}(50^{\circ}-10^{\circ})x\,cos\frac{1}{2}(50^{\circ}+10^{\circ})}{2sin\frac{1}{2}(50^{\circ}-10^{\circ})x\,sin\frac{1}{2}(50^{\circ}+10^{\circ})} = \cot30^{\circ} = -\sqrt{3}.$$

Equations and Identities

Referring to the figure on page 17, for any angle θ we have $ON^2 + OM^2 = OP^2$
Hence by the definitions:

$$\cos^2\theta + \sin^2\theta = 1.$$

Dividing by $\cos^2\theta$:

$$1 + \tan^2\theta = \sec^2\theta$$

Dividing by $\sin^2\theta$:

$$\cot^2\theta + 1 = \operatorname{cosec}^2\theta$$

These three identities and the identical relations on page 18 are used in the solution of trigonometric equations and the proof of other identities.
Many simple trigonometric equations are solved by **factorizing and equating the factors to zero.**

Example: Solve the equation $1 - \sin x = 2\cos^2 x$ for values of x between $0°$ and $360°$.

Substituting for $\cos^2 x$ and transferring to the L.H.S. $-2(1 - \sin^2 x) + 1 - \sin x = 2\sin^2 x - \sin x - 1 = 0$

$\therefore (2\sin x + 1)(\sin x - 1) = 0,$ $\qquad \therefore \sin x = -\frac{1}{2}$ or 1.

$\sin 30° = \frac{1}{2}$ $\quad \therefore$ if $\sin x = -\frac{1}{2}, x = 210°$ or $330°$.

$\sin 90° = 1,$ $\therefore$ if $\sin x = 1, x = 90°$.

Example: Solve $\sin 2x = \sin x$ for $0° \le x° \le 360°$

$\sin 2x = \sin x,$ $\therefore$ $2\sin x$ *times* $\cos x - \sin x = 0*$

$\therefore \sin x$ *times* $(2\cos x - 1) = 0.$ $\therefore \sin x = 0$ or $\cos x \frac{1}{2}$

$\therefore x = 0°, 180°, 360°,$ or $60°, 300°$.

*Written this way to avoid the possibility of dividing through by $\sin x$ which would **lose** the root $\sin x = 0$. The proof of identities involves **changing the L.H.S. into the R.H.S.** by use of the standard identities. **Examples.**

(i) $\dfrac{\sin A + \sin B}{\cos A + \cos B} = \dfrac{2\sin\frac{1}{2}(A+B) x \cos\frac{1}{2}(A-B)}{2\cos\frac{1}{2}(A+B) x \cos\frac{1}{2}(A-B)}$

$$= \frac{\sin\frac{1}{2}(A+B)}{\cos\frac{1}{2}(A+B)} = \tan\frac{1}{2}(A+B)$$

(ii) $(\sin A + \sin B)^2 + (\cos A + \cos B)^2 = \sin^2 A + \sin^2 B$
$+ 2\sin A\, x\, \sin B + \cos^2 A + \cos^2 B + 2\cos A\, x\, \cos B$

$= (\sin^2 A + \cos^2 A) + (\sin^2 B + \cos^2 B)$
$\qquad\qquad + 2(\cos A\, x\, \cos B + \sin A\, x\, \sin B)$
$= 2 + 2\cos(A - B).$

Solution of Triangles

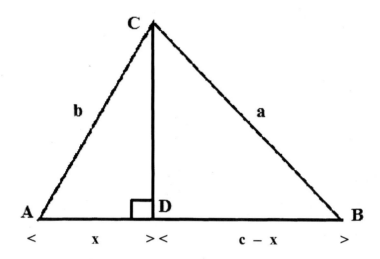

From the two right-angled triangles obtained by drawing the altitude through C, we have $CD = b \times \sin A = a \times \sin B$. Using the altitude through A gives $c \times \sin B = b \times \sin C$. These results are put together to give the **Sine Rule:**

$$\frac{a}{sinA} = \frac{b}{sinB} = \frac{c}{sinC}$$

Using Pythagoras in the two right-angled triangles:
$$b^2 = x^2 + CD^2 \quad \text{and} \quad a^2 = (c-x)^2 + CD^2$$
Subtracting: $a^2 - b^2 = c^2 - 2cx$ where $x = b \times \cos A$
Hence the **Cosine Rule: $a^2 = b^2 + c^2 - 2bc \times cosA$**

Similarly
$$b^2 = c^2 + a^2 - 2ca \times cosB$$
$$C^2 = a^2 + b^2 - 2ab \times cosC$$

Given **one side and two angles,** the third angle is found by the angle sum property and the other two sides by the Sine Rule.

Given **three sides,** find the largest angle (opposite the largest side) by the Cosine Rule, the second angle by the Sine Rule and the third angle from the angle sum property.

Given **two sides and the included angle,** find the third side by the Cosine Rule, the smaller unknown angle by the Sine Rule and the third angle from the angle sum. Do **not** find the larger angle first as the Sine Rule does not distinguish between acute and obtuse angles. Thus if A is the larger angle and

Solution of Triangles

$\sin A = 0.9231$, A is either $67° 23'$ or $112° 37'$.
Example: Find the remaining side and angles if $a = 4$ in, $c = 6$ in *and* $B = 40°$.

Cosine Rule: $\quad b^2 = c^2 + a^2 - 2ca \times \cos B$
$\qquad\qquad\qquad = 36 + 16 - 48 \times 0.7660$
$\therefore b = \sqrt{15 \times 23} = 3.903 = 3.90$ (by tables)
Smaller unknown angle is A (opposite smaller sides)

Sine Rule: $\qquad \dfrac{\sin A}{4} = \dfrac{\sin 40°}{3.903}$

$\therefore \sin A = 0.6589$ and $A = 41°13'$ (by tables)
$\therefore C = 180° - (40° + 41°14') = 98°47'$
Calculating C first, we get $\sin C = 0.9882$
$\therefore C = 81°13'$ or $98°47'$ and we are unable to decide which value to take.
When the calculation has to be done by logarithms the following alternative formulae are more suitable.

$$\cos\tfrac{1}{2}A = \sqrt{\frac{s(s-a)}{bc}} \qquad\qquad \sin\tfrac{1}{2}A = \sqrt{\frac{(s-b)(s-c)}{bc}}$$

$$\tan\tfrac{1}{2}A = \sqrt{\frac{(s-b)(s-c)}{s(s-a)}} \qquad\qquad \tan\tfrac{1}{2}(A-B) = \frac{a-b}{a+b} \times \cot\tfrac{1}{2}c$$

Where $s = \tfrac{1}{2}(a + b + c)$. The formulae for $\tfrac{1}{2}B, \tfrac{1}{2}C$ are obtained by changing the letters cyclically. The last formula is used when two sides and the included angle are given, for knowing a, b, and c the R.H.S. is easily computed by logarithms to give A-B. A +B is equal to $180° - C$ and A and B can then be found from the pair of simultaneous equations.

AREA OF TRIANGLE

The area of the triangle $= \dfrac{1}{2} \times AB \times CD = \dfrac{1}{2} \times c \times b\sin A$

$$= \tfrac{1}{2}bc \times \sin A$$
$$= \tfrac{1}{2}ca \times \sin B = \tfrac{1}{2}ab \times \sin C$$

Writing $\sin A$ as
$2\sin\tfrac{1}{2}A \times \cos\tfrac{1}{2}A$ and using the formulae for $\sin\tfrac{1}{2}A$ and $\cos\tfrac{1}{2}A$:
The area of the triangle $= \sqrt{s(s-a)(s-b)(s-c)}$
The first formula is used when two sides and the included angle are known and the second formula when three sides are known.

3-Dimensional Geometry

A **plane** is a surface in which the straight line joining any two points in the surface lies entirely in the surface. A line not in the plane meets the, plane in one point unless it is parallel to the plane. If the line is **perpendicular to every straight line in the plane** through the point of intersection, the line is a **normal** to the plane.

The foot of the normal from a point to a plane is the **projection** of the point on the plane. The projections of the points of a straight line on to a plane are the points of a straight line which is the **projection of the first line on the plane.**

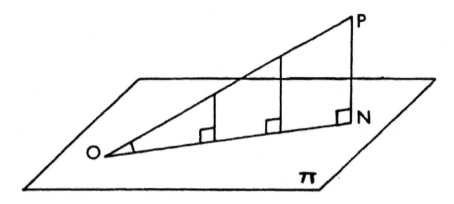

The **angle between a straight line and a plane** is the **angle between the line and its projection on the plane**. Angle *PON* is the angle between *OP* and the plane π.

SKEW LINES

Lines which are **not parallel** and which **do not meet** are skew. The angle between two skew lines is the angle between intersecting lines parallel to the skew lines. The distance of a point from a line or a plane is the perpendicular distance which is also the **shortest distance.** The shortest distance between two skew lines is the length of the line perpendicular to each of the lines, and equals the distance of any point on one from the plane through the second parallel to the first.

Example: The ends *ABCD, EFGH* of a cuboid of length 4 are in squares of side 3 inches. Find the shortest distance between the edge *CD* and the diagonal *AG.*

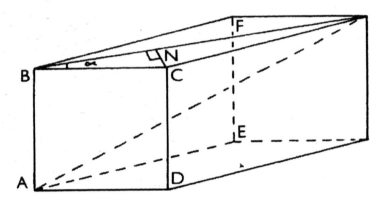

AG and *CD* are skew and the plane *ABG* containing *AG* is parallel to CD. Hence the shortest distance between *AG* and *CD* equals the perpendicular distance of C from plane *ABG*, **i_e.** *CN* = *BC* sin α. Triangle *BCG* is a 3, 4, 5 triangle so sin α = 0.8 and *CN* = 3 x 0.8 = 2.4 in.

ANGLE BETWEEN TWO PLANES

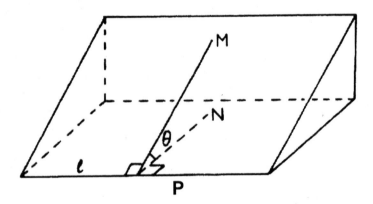

Two non-parallel planes intersect in a straight line *l*. *P* is any point on *l* and *PM, PN* are straight lines, **one in each plane, perpendicular** to *l*. The angle **MPN between these lines is the angle between the planes.**
The **line of greatest slope** in a plane is a line **perpendicular** to the line of intersection of the plane and the **horizontal** plane.

3-Dimensional Trigonometry

The geometry of the last two pages is required mainly in the solution of problems in trigonometry. When solving such problems, **draw a clear diagram,** pick out the **appropriate triangles** and **draw separately;** find the sides and angles of these triangles as necessary.

Example. A pyramid *OABCD* has a square base *A BCD* of side 6 in and vertex 0, 5 in from each corner of the base. Calculate (i) the angle made by *OA* with the base, (ii) the angle between the planes *OAB* and *OBC*.

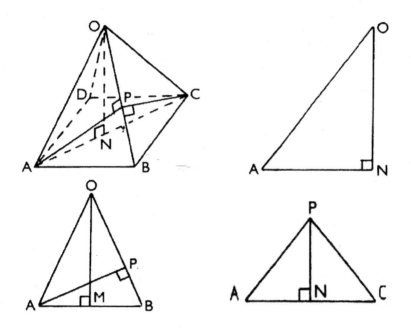

(i)   The projection of *OA* on the base is *AN* where *N* is the midpoint of *AC*, so required angle is ∠*OAN*. *OA* = 5, *AN* = 3 ⎺, so cos ∠*OAN* = 0·8484 and ∠*OAN* = 31° 58'.

(ii)   Take *P* on *OB*, the intersection of the planes, so that *AP, CP* are each perpendicular to *OB*. Required angle is ∠*APC*. Δ*BMO* is a 3, 4, 5 Δ so sin α = 0.8.

In Δ *ABP*, *AP* = *AB* sin α = 6 x 0.8 = 4.8.

In Δ *APN*, sin ∠*APN* = *AN/AP* = 0·8837 and ∠*APN* = 62° 6'.

Δ A*PC* is isosceles so ∠ *APC* = 2. ∠*APN* = 124° 12'.

Points and Loci

Any **point** in a plane is defined by the **ordered pair** of numbers (x, y), the measures of its distances from the intersecting perpendicular axes, YOY', XOX'. XOX' is the **x-axis**, YOY' is the **y-axis**, 0 is the **origin**, (x, y) are the **coordinates** of the point. X is positive or negative according as it is measured in the X or X' direction, and a similar rule holds for y.

Let A (x_1, y_1) and B (x_2, y_2) be any two points and draw through A and B lines parallel to the axes. (Fig. 1)

$AC = x_2 - x_1$, $CB = y_2 - y_1$, and $AB^2 = AC^2 + CB^2$

∴ **Distance AB** $= \sqrt{(x2 - x1)2 + (y2 - y1)2}$

Example: The distance between the points (-1, 2) and (3,5) =

$$\sqrt{\{3 - (-1)\}2 + (5 - 2)^2} = \sqrt{16 + 9} = \sqrt{25} = 5.$$

If P is the **midpoint** of AB and the parallels to the axes through P meet $4C$ and BC in M and N respectively, then M and N are the midpoints of AC and CB. (Fig. 1)

∴ x coordinate of $P = x_1 + \frac{1}{2}(x_2 - x_1) = \frac{1}{2}(x_1 + x_2)$ and

y coordinate of $P = y_1 + \frac{1}{2}(y_2 - y_1) = \frac{1}{2}(y_1 + y_2)$

Generally, if P divides AB internally or externally in a given ratio, M and N divide AC and CB in the same ratio; the coordinates of P can then be found by the same method.

Example: If $AP:PB = 1:2$, then $AP = \frac{1}{3}AB$, $AM = \frac{1}{3}AC$, $CN = \frac{1}{3}CB$ and x coordinate of $P = x_1 + \frac{1}{3}(x_2 - x_1)$, etc.

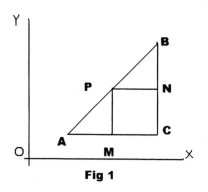

Fig 1

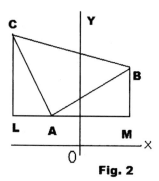

Fig. 2

The **area of a triangle** with given vertices can always be found in terms of the areas of a **trapezium and two right-angled triangles.**

Example: Find the area of the triangle whose vertices are A (-1, 1), B (2, 3) and C (—2, 5).
Draw a parallel to the x-axis through the lowest vertex and parallels to the y-axis through the other vertices to meet the first line in L and M. (Fig. 2)

Area $\triangle ABC$ = area trap. $BCLM$ - *area* $\triangle AMB$ - area $\triangle ACL$

$$= \tfrac{1}{2}(CL + BM) \text{ x. } LM - \tfrac{1}{2}.x \; AM \text{ x } MB - \tfrac{1}{2}AL \text{ x } LC$$

$$= \tfrac{1}{2}(4+2) \text{ x } 4 - \tfrac{1}{2} \text{ x } 3 \text{ x } 2 - \tfrac{1}{2} \text{ x } 1 \text{ x } 4 = 7 \text{ unit}^2$$

The **gradient** of a line is defined to be the **tangent of the angle** the line makes with the **positive** direction of the x-axis. Referring to Fig 1.

$$\text{Gradient AB} = \tan\angle\text{BAC} = \frac{BC}{AC} = \frac{y_2 - y_1}{x_2 - x_1}$$

A **locus** is a set of points satisfying given conditions and is obtained as a **relation** between the x and y coordinates of any point satisfying the conditions. This relation is called the **equation of the locus.**

Example: Find the equation of the locus of points equidistant from the points O (0, 0) and A (1, 1).
Let P, coordinates *(h, k)*, be any point on the locus.
The given condition is that $OP = AP$, $\therefore OP^2 = AP^2$
$\therefore (h — 0)^2 + (k — 0)^2 = (h — 1)^2 + (k — 1)2$
which when simplified gives $h + k = 1$. This is a relation between the x and y coordinates of any point on the locus, so the equation of the locus is

$$x + y = 1.$$

The point $(2, — 1)$ is on the locus since $2 + (— 1) = 1$.

If two curves (loci) **intersect,** the coordinates of the point(s) of intersection **must satisfy** the equations of both curves. To find the coordinates of the point(s) we solve a pair of simultaneous equations.

Example: The curves whose equations are $y = x + 2$ and $y = x^2$ intersect where $x^2 = x + 2$ or $x^2 — x - 2 = 0$. The roots of this equation are $x = 2$ and $x = — 1$. When $x = 2$, $y = 4$ and when $x = —1, y = 1$.
$\therefore$ the points of intersection are (2, 4) and (— 1, 1).

Straight Lines

A **straight line** has a **constant gradient (m)** and makes an **intercept (C)** on the y-axis. If $P (x, y)$ is any point on the line the gradient $m = \frac{y-c}{x}$ and hence

$$Y = mx + C$$

which is the equation of the line.

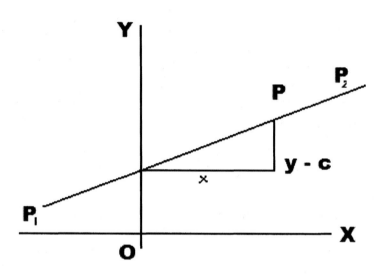

Any equation of the form **ax + by + c = 0** can be written $y = -\frac{a}{b}x - \frac{c}{b}$ and thus is the **equation of a straight line with gradient** $-\frac{a}{b}$

Thus $x - 2y - 4 = 0$ can be written $y = \frac{1}{2}x - 2$ and so represents a straight line gradient $\frac{1}{2}$ and intercept on the y-axis $= -2$. If the line goes through the point P_i (x_i, y_i), $y_i = mx_1 + c$. Subtracting from $y = mx + c$ we obtain

$$y - y_1 = m(x - x_1)$$

as the equation of the straight line of **gradient m through the point (x_1, y_1).**

Example: The line of gradient $\frac{2}{3}$ which goes through **(2, —1) has equation**
$$y - (-1) = \frac{2}{3}(x - 2)$$
i.e. **3y + 3 = 2x — 4 or 2x — 3y — 7 = 0**

If the line goes through P_1 (x_1, y_1) and $P_2(x_2, y_2)$, $m = \frac{y_2 - y_1}{x_2 - x_1}$. Substituting into $y - y_1 = m(x - x_1)$ and rearranging, the equation of the straight line joining (x_1, y_1) and (x_2, y_2) is:

$$\frac{y - y_1}{x - x_1} = \frac{y_2 - y_1}{x_2 - x_1}$$

Example: The equation of the line joining (2, 3) and (1, 4) is $\frac{y-3}{x-2} = \frac{4-3}{1-2} = -1$,

i.e. $y = -3 = -x + 2$

i.e. $x + y = 5$.

Two lines of gradients m_1, m_2 are **parallel** only if $m_1 = m_2$. Any line parallel to $ax + by + c = 0$ must be of the form $ax + by + c' = 0$ to satisfy this condition.

The lines are **perpendicular** only if the angles they make with the positive x-axis, α and β, differ by 90°.

$\beta - \alpha = 90^0$; $m_1 = \tan\alpha$; $m_2 = \tan\beta = \tan(\alpha + 90°)$

$$= \frac{\sin(\alpha + 90°)}{\cos(\alpha + 90°)} = \frac{\cos\alpha}{-\sin\alpha} = \frac{1}{\tan\alpha} = \frac{1}{m_1}$$

So $m_1 m_2 = -1$, i.e. **the product of the gradients** $= -1$. Any line perpendicular to $ax + by + c = 0$ must be of the form $bx - ay + c' = 0$ to satisfy this condition.

Example: The join of A (0, 3) and B(2, 4) has gradient $= \frac{1}{2}$ so the gradient of the perpendicular to $AB = -2$.

Midpoint of AB is $\{\frac{1}{2}(0 + 2), \frac{1}{2}(3 + 4)\} = (1, 3\frac{1}{2})$.

$\therefore$ the equation of the perpendicular bisector of AB is

$$y - 3\frac{1}{2} = -2(x - 1), \text{ i.e. } 4x + 2y = 11.$$

Generally, if θ is the angle between the lines,

$\theta = \beta - \alpha$ and $\tan\theta = \frac{\tan\beta - \tan\alpha}{1 + \tan\beta \times \tan\alpha} = \frac{m_2 - m_1}{1 + m_2 m_1}$

The **perpendicular distance of the point** (x_1, y_1) from the line $ax + by + c = 0$ is :

$\pm \dfrac{ax_1 + by_1 + c;}{\sqrt{a^2 + b^2}}$ the sign is taken to make the distance of the origin positive.

The Circle

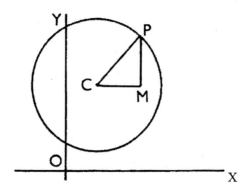

To find the equation of the circle, center C (1, 3) and radius 2, let $P(x, y)$ be any point on the circle.

$CM^2 + MP^2 = CP^2$ where $CM = x - 1$, $MP = y - 3$, $CP = 2$ $\therefore$ $(x - 1)^2 + (y - 3)^2 = 2^2$ which simplifies to give $x^2 + y^2 - 2x - 6y + 6 = 0$ as the required equation. Generally, if **C is ($-g$, $-f$), r the radius** we obtain the equation
$$x^2 + y^2 + 2gx + 2fy + c = 0$$
where $C = g^2 + f^2 - r^2$.

Any **second** degree equation in which **(1) the coefficients of x^2 and y^2 are equal, (2) there is no xy term** can be written in the form $x^2 + y^2 + 2gx + 2fy + c = 0$ by dividing by the coefficient of x^2 and y^2, if it is not 1, and is, therefore a **circle, centre ($-g$, $-f$)** and **radius** $\sqrt{g + f^2 - c}$.

To find the equation of the circle through **three given points,** substitute the coordinates of the points in turn in $x^2 + y^2 + 2gx + 2fy + c = 0$ and solve the three resulting simultaneous equations for g, f and c.

To find the equation of the **tangent** to a circle at the point $P(x_1, y_1)$:
    (1) find the coordinates of the centre C:
    (2) write down the gradient of the radius CP using the points C and P
    (3) find the gradient of the tangent using $m_1 m_2 = -1$
    (4) Substitute in $y - y_1 = m(x - x_1)$.

Gradients and Differentiation

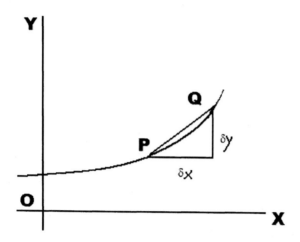

The **gradient of a curve** at a point P is the gradient of the tangent at P. If P is (x, y) and a nearby point Q is $(x + \delta x, y + \delta y)$ the gradient of the chord $PQ = \frac{\delta y}{\delta x}$
The **tangent** at P is the **limiting position of chord PQ** as Q approaches P, i.e. as $\delta x \to 0$, so the gradient at P
$= \lim (\text{gradient of chord PQ}) = \lim \frac{\delta y}{\delta x}$ and written $\frac{dy}{dx}$
and called the **derivative of y with respect to x.** y is a function of x and the process of finding the derivative of a function is called **differentiation.** The method of finding a derivative from **first principles** is shown in the differentiation of $y = x^2$.

Let δx, δy be corresponding increments in x and y. Corresponding values of x and y are related by $y = x^2$
$\therefore y + \delta y = (x + \delta x)^2 = x^2 + 2x \cdot \delta x + (\delta x)^2$
But $y = x^2$ so $\delta y = 2x \cdot \delta x + (\delta x)^2$ and $\frac{\delta y}{\delta x} = 2x + \delta x$.
$\therefore \frac{dy}{dx} = \lim \frac{\delta y}{\delta x} = \lim (2x + \delta x) = 2x$.
The gradient of the curve at $(2,4) = 2 \cdot 2 = 4$

Negative and fractional powers of x are dealt with similarly and lead to the formulation of the **rule** for the **differentiation of powers of x:**

$$\text{If } y = x^n, \frac{dy}{dx} = n \cdot x^{n-1}$$

DIFFERENTIATION OF A CONSTANT

A constant coefficient is unchanged by differentiation so

if $y = A \cdot x^n$, $\dfrac{dy}{dx} = A \cdot nx^{n-1}$

If y = a constant, a change δx in x causes no change in

y so $\delta y = 0$, $\dfrac{\delta y}{\delta x} = 0$, hence $\dfrac{dy}{dx} = 0$ **if y is constant.**

TRIGONOMETRIC FUNCTIONS

If $y = \sin x$ and δx, δy are corresponding increments in x and y, $y + \delta y = \sin (x + \delta x)$ and $y = \sin x$.

$\therefore \delta y = \sin(x + \delta x) - \sin x = 2 \sin\tfrac{1}{2} \delta x \cdot \cos(x + \tfrac{1}{2}\delta x)$

$\therefore \dfrac{dy}{dx} = \lim \dfrac{\delta y}{\delta x} = \lim \dfrac{\sin(\frac{\delta y x}{2})}{\delta x/2} \cdot \cos(x + \delta x/2)$

$\quad \bullet \qquad = 1 \cdot \cos x = \cos x$

Hence $\dfrac{d}{dx}\sin x = \cos x$. Similarly, $\dfrac{d}{dx}\cos x = -\sin x$ and $\dfrac{d}{dx}\tan x = \sec^2 x$

DIFFERENTIATION OF A SUM

If $y = u + v + ...$ where u, v, ... are functions of x and δx, δy, δu, δv, ... are corresponding increments, then $y + \delta y = (u + \delta u) + (v + \delta v) + ...$

$\therefore \delta y = \delta u + \delta v + ...;$ hence, dividing by δx and then letting

$$\delta x \to 0, \dfrac{dy}{dx} = \dfrac{du}{dx} + \dfrac{dv}{dx} + ...$$

Thus the **derivative of a sum is equal to the sum of the derivatives of the separate terms.**

e.g. $\dfrac{d}{dx}(x^3 - 3x^2) = \dfrac{d}{dx}(x^3) - \dfrac{d}{dx}(3x^2) = 3x^2 - 6x$

DIFFERENTIATION OF A PRODUCT

If $y = u \cdot v$, $y + \delta y = (u + \delta u)(v + \delta v)$. Expand $y + \delta y$, subtract $y = u \cdot v$ and divide by δx to obtain

$\dfrac{\delta y}{\delta x} = u \cdot \dfrac{\delta v}{\delta x} + v \cdot \dfrac{\delta u}{\delta x} + \delta u \cdot \dfrac{\delta v}{\delta x}$. When $\delta x \to 0$, $\delta u \to 0$, hence

$$\dfrac{d}{dx}(u \cdot v) = u \cdot \dfrac{dv}{dx} + v \cdot \dfrac{du}{dx}$$

DIFFERENTIATION OF A QUOTIENT

If $y = \dfrac{u}{v}$, $y + \delta y = \dfrac{u + \delta u}{v + \delta v}$ and $\delta y = \dfrac{u + \delta u}{v + \delta v} - \dfrac{u}{v}$

$= \dfrac{(u + \delta u).v - (v + \delta v).u}{(v + \delta v).v}$ Hence $\dfrac{\delta y}{\delta x} = \dfrac{v.\dfrac{\delta u}{\delta x} - u.\dfrac{\delta v}{\delta 8 x}}{v^2 + v.\delta v}$

When $\delta x \to 0$, $\delta v \to 0$, hence

$$\frac{d}{dx}\left(\frac{u}{v}\right) = \frac{v.\dfrac{du}{dx} - u.\dfrac{dv}{dx}}{v^2}$$

Example: Differentiate (i) $x . \sin x$, (ii) $\tan x$.

(i) Let $y = x . \sin x$, $u = x$, $v = \sin x$ and $y = u .v$

$\dfrac{du}{dx} = 1$, $\dfrac{dv}{dx} = \cos x$ and $\dfrac{dy}{dx}$

(ii) $= u . \dfrac{dv}{dx} + v . \dfrac{du}{dx}$

$= x . \cos x + \sin x.$

(iii) Let $y = \tan x = \dfrac{\sin x}{\cos x}$ and $u = \sin x$, $v = \cos x$.

Then $\dfrac{du}{dx} = \cos x$ and $\dfrac{dv}{dx} = -\sin x$.

$\therefore \dfrac{dy}{dx} = \dfrac{\cos x . \cos x - \sin x .(-\sin x)}{\cos^2 x} = \dfrac{\cos^2 x + \sin^2 x}{\cos^2 x}$

$= \sec^2 x$ since $\cos^2 x + \sin^2 x = 1$.

FUNCTION OF A FUNCTION

When y is some function of u and u is some function of x, a change δx in x causes a change δu in u, which in turn causes a change δy in y.

By the ordinary rules for fractions, $\dfrac{\delta y}{\delta x} = \dfrac{\delta y}{\delta u} . \dfrac{\delta u}{\delta x}$

When $\delta x \to 0$, $\delta u \to 0$, hence

$$\frac{dy}{dx} = \frac{\delta y}{\delta u} . \frac{\delta u}{\delta x}$$

The function of a function rule is particularly useful when dealing with **powers of a function of x.**

Example: Differentiate $(x^2 + 5)^7$ with respect to x.

Ley $y = (x^2 + 5)^7$, $u = x^2 + 5$, so $y = u^7$.

$\dfrac{dy}{du} = 7u^6$, $\dfrac{du}{dx} = 2x$ and $\dfrac{dy}{dx} = \dfrac{dy\,du}{du\,dx} = 7u^6 . 2x = 14x(x^2 + 5)^6$

Tangents and Normals

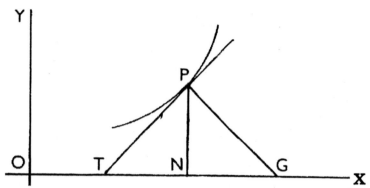

The **normal** at any point on a curve is the line through the point **perpendicular to the tangent**. The value of the derivative at the point (x_1, y_1) is the gradient of the tangent; the gradient of the normal is found from $m_1\, m_2 = -1$. The **equations** of the tangent and normal at (x_1, y_1) are then found by using $y - y_1 = m(x - x_1)$.

Example: The point (2, 4) is on the curve $y = x^3 - x^2$

$\dfrac{dy}{dx} = 3x^2 - 2x$ so the gradient of the tangent at (2,4) is $3 \cdot 2^2 - 2 \cdot 2 = 8$ and the

gradient of the normal is $-\dfrac{1}{8}$

∴ Equation of tangent is $y - 4 = 8(x - 2)$

i.e. $y = 8x - 12$.

Equation of normal is $y - 4 = \dfrac{1}{8} \text{-}(x - 2)$

i.e. $8y - 32 = -x + 2$ or $x + 8y - 34 = 0$.

SUBTANGENT AND SUBNORMAL

Let the tangent, normal and ordinate at $P(x, y)$ meet the x-axis in T, G and N respectively. The length TN is the **subtangent** and the length NG is the **subnormal**.

Gradient of tangent $PT = \dfrac{PN}{TN} = \dfrac{dy}{dx}$

$\therefore$ **TN = length of subtangent = y $\div \dfrac{dy}{dx}$**

$NG = PN \times \tan \angle NPG = y \times \tan \angle PTN$

$\therefore$ **NG = length of subnormal = y $\times \dfrac{dy}{dx}$**

Velocity and Acceleration

The distance of a body from an origin is a function of the time t. If δx, δt are corresponding increments in x, t the average rate of change of distance, i.e. the average velocity, is $\frac{\delta x}{\delta t}$. The $\lim \frac{\delta x}{\delta t}$, i.e. $\frac{dx}{dt}$ is defined as the **velocity at time t.**

Example.: If the distance x meter after t sec is given by

$$x = \frac{1}{2}t^2 + 4t, \ \frac{dx}{dt} = t + 4 = 6 \text{ when } t = 2.$$

$$\therefore \text{ velocity after 2 sec} = 6 \text{ m/s.}$$

Acceleration is the **rate of change of velocity v** so $\frac{dv}{dt}$ is defined as the **acceleration at time t.**

$v = \frac{dx}{dt}$, so acceleration $= \frac{d}{dt}(\frac{dx}{dt})$ written $\frac{d^2x}{dt^2}$, the second derivative of x with respect to t.

In the example above, $\frac{dx}{dt} = t + 4$, $\therefore \frac{d^2x}{dt^2} = 1$

i.e. the acceleration is constant $= 1$ m/s^2.
By the function of a function rule $\frac{dv}{dt} = \frac{dv}{dx} \cdot \frac{dx}{dt} = \frac{dv}{dx} \cdot v$

Thus the acceleration is also $v\frac{dv}{dx}$

The function of a function rule may be used to solve physical problems involving a rate of increase.

Example: The radius (r) of a soap bubble is 5 cm and is increasing at the rate of 0.2 cm/s. Find the rate of change of its volume (V) at this time (t).

$$V = \frac{4}{3}\pi r^3, \ \therefore \frac{dV}{dr} = 4\pi r^2 \cdot \frac{dr}{dt} = 0.2 \text{ cm/s}$$

$$\frac{dV}{dt} = \frac{dV}{dr} \cdot \frac{dr}{dt} = 4\pi r^2 \cdot (0.2) = 4\pi \cdot 25 \cdot (0.2) \text{ when } r = 5$$
$$= 20\pi \text{ cm}^3/\text{s.}$$

In general, if the variable y is a function of the variable x, $\frac{dy}{dx}$ gives the **rate of change of y with respect to x.**

y increases or decreases as x increases according as $\frac{dy}{dr}$ is positive or negative.

Maxima and Minima

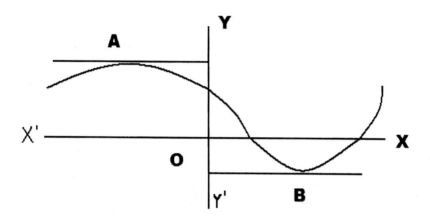

The diagram represents part of the graph of $y = f(x)$. A and B are **turning points** of the curve. The ordinate at A is **greater** than any neighboring ordinate and A is called a **maximum point**. The ordinate at B is **less** than any neighboring ordinate and B is a **minimum point**. At both A and B the **tangent** to the curve is **parallel** to the x-axis and so $\frac{dy}{dx} = 0$ **at these points.**

In passing through A in the sense of x increasing the gradient changes from positive to negative, i.e. it is decreasing. Thus the rate of change of the gradient $= \frac{d}{dx}(\frac{dy}{dx}) = \frac{d^2y}{dx^2}$ is negative at A.

At a maximum turning point, (I) $\frac{dy}{dx} = 0$

(II) $\frac{d^2y}{dx^2} < 0$

In passing through B in the sense of x increasing the gradient changes from negative to positive, i.e. it is increasing. Thus $\frac{d}{dx}(\frac{dy}{dx}) = \frac{d^2y}{dx^2}$ is positive at B.

At a minimum turning point, (I) $\frac{dy}{dx} = 0$

(II) $\frac{d^2y}{dx^2} > 0$

Example: Find the turning points of the curve
$$y = 2x^3 + 3x^2 - 12x + 6$$
and discriminate between them (the turning points).
$\frac{dy}{dx} = 6x^2 + 6x - 12 = 6(x + 2)(x - 1) = 0$ when $x = -2$ or 1 (and corresponding $y =$ 26 or -1)
∴ the turning points are (-2, 26) and (1, -1).
$\frac{d^2y}{dx^2} = 12x + 6 = -18$ when $x = -2$; $= +18$ when $x=1$.
∴(-2, 26) is a maximum turning point
And (1, -1) is a minimum turning point.
The **same conditions** are used to determine the maximum and minimum values
of a **function** of a given variable.

PRACTICAL PROBLEMS

A number of practical problems can be solved by the above methods. The
quantity whose maximum value is to be determined has often to be expressed
initially **in terms of two variables. A geometrical or other relation** between
the variables has then to be sought in order to eliminate one of them. In many
cases it is obvious from physical considerations whether a maximum or a
minimum value has been obtained and it is unnecessary to consider the second
derivative.

Example: An open rectangular box with a square base has to have a volume of
500 cm³. Find the minimum area of sheet metal used in the box.
Let A cm² be the surface area of the box.

  x cm be the length of the side of the square base,
  y cm be the height of the box.

$A = x^2 + 4xy$. Volume of box $= x^2y = 500$ cm³. Hence substituting $y =$
$500/x^2$ in the formula for A

$$A = x^2 + \frac{2000}{x} \text{ and } \frac{dA}{dx} = 2x - \frac{2000}{x^2} = \frac{2(x^3 - 1000)}{x^2}$$

$= 0$ for a maximum or minimum value of A.
∴ $x^3 - 1000 = 0$, i.e. $x = 10$
When $x = 10$, $y = 5$ and $A = 100 + 200 = 300$
$\frac{d^2A}{dx^2} = 2 + \frac{4000}{x^3} > 0$ for all possible values of x.
i.e. A is a minimum.
∴ minimum area of sheet metal used $= 300$ cm² when the dimensions of the
box are 10 cm x 10 cm x 5 cm.

Integration

Integration is the **inverse** of differentiation and involves finding a **function whose derivative is known.**

Thus, if $\frac{dy}{dx} = 2x$ we write $y = \int 2x \cdot dx = x^2 + c$ since the derivative of x^2 is $2x$ and the derivative of any constant is zero. The constant c is **arbitrary** and cannot be determined without more information; e.g. given that $y = 0$ when $x = 1$ it follows that $c = -1$.

$\int 2x \cdot dx$ is called the **indefinite integral** of $2x$.

Integrals have to be recognized as the results of particular differentiations and the following standard integrals must be remembered:

Since $\frac{d}{dx}x^n = n \cdot x^{n-1}$, $\int x^n \cdot dx = \frac{x^{n+1}}{n+1} + c \ (n \neq -1)$

Since $\frac{d}{dx}\sin x = \cos x$, $\int \cos x \cdot dx = \sin x + c$

Since $\frac{d}{dx}\cos x = -\sin x$, $\int \sin x \cdot dx = -\cos x + c$

The arbitrary constant must **not** be omitted from an indefinite integration and the answer when differentiated must give the function that was integrated.
A constant coefficient is not altered in integration. A **constant A standing alone integrates into Ax + c.**
Since the derivative of a sum is the sum of the derivatives it follows that the **integral of a sum is the sum of the integrals of the separate terms.**
There are no rules for the integration of products and quotients and they have to be **multiplied or divided out** and then integrated as a sum of terms.

Example: $\int x(x^2 - 2) \cdot dx = \int (x^3 - 2x) \cdot dx$

$= \int x^3 \cdot dx - \int 2x \cdot dx = \frac{x^4}{4} - 2 \cdot \frac{x^2}{2} + c = \frac{1}{4}x^4 - x^2 + c$

Example: $\int \{(x^4 - x^2 + 1)/x^2\} \cdot dx = \int (x^2 - 1 + x^{-2}) \cdot dx$

$= \frac{x^3}{3} - x + \frac{x^{-1}}{-1} + c = \frac{1}{3}x^3 - x - \frac{1}{x} + c.$

Some integrals require the use of the **function of a function rule in reverse** and the integration is effected by means of a **substitution.**

Example: If $y = \int x(x^2 + 3)^3 \cdot dx$ then $\frac{dy}{dx} = x(x^2 + 3)^3$

$\frac{dy}{dx} = \frac{dy}{du} \cdot \frac{du}{dx}$ where $u = x^2 + 3$ and $\frac{du}{dx} = 2x$

$\therefore \frac{dy}{du} \cdot 2x = x(x^2 + 3)^3 = xu^3$ and hence $\frac{dy}{du} = \frac{1}{2}u^3$

$\therefore y = \int \frac{1}{2}u^3 \cdot du = \frac{1}{2} \cdot \frac{u^4}{4} + c = \frac{1}{8}(x^2 + 3)^4 + c$

Here, u has replaced a function of x; the last example below shows x replaced by a function of u.

DEFINITE INTEGRAL

The symbol $\int_2^1 f(x) \cdot dx$ is used to represent the **value of the integral when x = b minus its value when x = a**. It is called the **definite integral** and is expressed symbolically as [G(x)] where $G(x)$ is a function whose derivative is $f(x)$, **a and b are called the limits of integration.** The arbitrary constant vanishes on subtraction and so is **omitted** in definite integration.

Example: $\int_1^2 (x^2 - 1) \cdot dx = \left[\frac{x^3}{3} - x\right]$

$= (\frac{8}{3} - 2) - (\frac{1}{3} - 1) = \frac{2}{3} + \frac{2}{3} = 1\frac{1}{3}.$

A definite integral evaluated by substitution has to have its **limits changed** to those for the new variable.

Example: Let $y = \int_0^1 \frac{1}{\sqrt{1-x^2}} \cdot dx$ and put $x = sin\ u$

$\frac{dy}{du} = \frac{dy}{dx} \cdot \frac{dx}{du} = \frac{1}{\sqrt{1-sin^2u}} \cdot cos\ u = \frac{1}{cos\ u} \cdot cos\ u = 1$

When x = 0, u = 0; when x= 1, $u = \frac{\pi}{2}$

$\therefore y = \int_0^{\frac{\pi}{2}} 1 \cdot du = [u]$

$= \frac{\pi}{2} - 0 = \frac{\pi}{2}$

Areas and Volumes

AREA UNDER A CURVE

The **area** bounded by the curve $y = f(x)$, the x-axis and the ordinates $x = a$ and $x = b$ is given by the **definite integral**

$\int_a^b y \, dx$ **where** $f(x)$ **replaces** y **for evaluation.**

If the curve cuts the x-axis at $x = c$ between $x = a$ and $x = b$, the area is the **sum** of the **numerical** values of $\int_a^c y \, dx$ and $\int_c^b y \, dx$, the area for the part below the x-axis being **negative.**

Example: Find the area of the segment of the curve $y = x(3 - x)$ cut off by the x-axis.

The curve cuts the x-axis at $x = 0$ and $x = 3$.

$\therefore$ Area of segment $= \int_0^3 (3x - x^2) \, dx = [\frac{3x^2}{2} - \frac{x^3}{3}]_0^3$

$= (13\frac{1}{2} - 9) - (0 - 0) = 4\frac{1}{2}$

AREA BETWEEN CURVES

To find the area **between** two curves, find the **x coordinates of the points of intersection** for the **limits of integration,** then the area under each curve **separately** and **subtract.**

VOLUME OF REVOLUTION

Revolution of the area through four right angles about the **x-axis** generates a solid whose **volume** is given by the **definite integral** $\pi \int_a^b y^2 \, dx$. If an area is rotated about the y-axis, interchange x and y and use the limits for y in the integral.

Example: The part of the curve $y = x^2 - 1$ between $y = 1$ and $y = 2$ is revolved about the y-axis.

Volume of solid formed $= \pi \int_1^2 x^2 \, dy = \pi \int_1^2 (y + 1) \, dy$

$= \pi[\frac{y^2}{2} + y]_1^2 = \pi(2 + 2) - \pi(\frac{1}{2} + 1) = \frac{5}{2}\pi$

Forces and Moments

Forces are of three kinds: (1) **attractions,** e.g. the weight of a body due to the earth's attraction; (2) **tensions** or **thrusts,** as in taut strings and rods; (3) **reactions,** the equal and opposite forces of bodies in contact. A force is described by giving its **magnitude, direction** and **line of action.** In SI units, a force is measured in **Newtons,** symbol N.

A force can be considered as acting at **any** point on its line of action. Two forces acting on a body maintain equilibrium only if they are **equal in magnitude, opposite in direction and have the same line of action.**

MOMENTS

The **moment** of a force about a **point** (or **axis**) is the **product of the force and the perpendicular distance of its line of action from the point. The Principle of Moments** states that **if a body is in equilibrium the sums of the clockwise and anti-clockwise moments about the same point are equal.**

Example: A light rod AB, 6 m long, is supported at its ends. Forces of 5, 10 and 8 N act vertically downward at 1, 3 and 5 m from A. Find the reactions at A and B.

Let the reactions at A and B be P N and Q N.
Clockwise moment about A = 5 x 1+ 10 x 3 +8 x 5 N m. Anti-clockwise moment about A= Q x 6 N m.
For equilibrium: 6Q = 75 and Q = 12.5.
Similarly, moments about B gives P = 10.5.
∴ reactions at A and B are 10˙5 N and 12˙5 N.

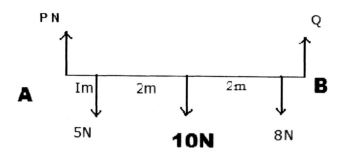

Vector Quantities

Quantities such as mass, time, speed requiring **only a magnitude** for their specification are called **scalar quantities.** Quantities like displacement, velocity, acceleration, force require **a direction as well as a magnitude** and are called **vector quantities.** We shall concentrate on forces but the methods apply to all vector quantities.

Concurrent forces P and Q whose lines of action are inclined at an angle α to each other are equivalent to a single force obtained by drawing a **force diagram.**

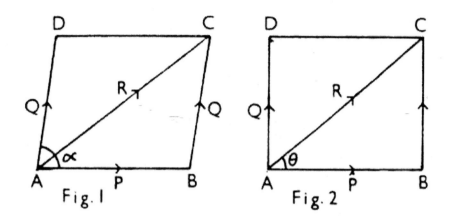

Fig. I   Fig. 2

AB - AD are drawn to **scale** to represent P and Q in **both magnitude and direction.** The parallelogram $ABCD$ is completed by drawing the parallels through B and D.

AC represents the single force R equivalent to P and Q. R is called the **resultant** of P and Q, and P and Q are the **components** of R. (Fig.

RESOLUTION OF FORCES

If the forces P and Q are at **right angles** the force diagram is a **rectangle.** (Fig. 2.) If 0 is the angle made by R with the force P,

$$R = \sqrt{(P^2 + Q^2)} \text{ and } \tan\theta = Q/P.$$

Also, $P = R\cos\theta$ and $Q = R\sin\theta = R\cos(90° - \theta).$

P and Q are the **resolved parts** or **resolutes** of R in the directions AB and AD.

N.B. The resolute of F in a direction at angle θ with that of F is $F\cos\theta$.

Hence any force can be resolved into two resolutes at right angles to each other. Conversely, if the resolutes are known the magnitude and direction of the resultant force can be calculated.

The **resultant of a system of forces** can be found as follows:

1. **Resolve each force into its resolutes in two mutually perpendicular directions.**

2. **Find the resultant resolute in each direction.**

3. **Calculate the resultant of the resultant resolutes.**

Example: Forces of 2, 4 and 5 N act at a point *0*. The first and second forces are inclined at 60° and 120° to the third force. Find their resultant.

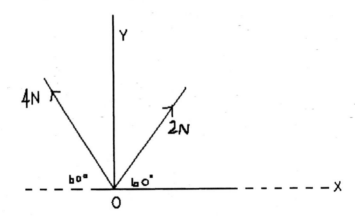

Take *OX* in the direction of the 5 N force and *OY* perpendicular to *OX*.
Resolve each force in the directions *OX, OY*.
The resolved parts of 5 N are (5, 0).
The resolved parts of 2 N are (2 cos 60°, 2 sin 60°). The resolved parts of 4 N are (4 cos 120°, 4 sin 120°). If *P, Q* are the resultant resolutes parallel to *OX, OY*
$P = 5 + 2 \cos 60° — 4 \cos 60° = 4$.
$Q = 0 + 2 \sin 60° + 4 \sin 60° = 3\sqrt{3}$.
The three forces are therefore equivalent to forces of 4 and $3\sqrt{3}$ N along *OX* and *0* Y.
Hence if *R* N is the resultant at angle θ with *OX*,
$R = \sqrt{(P^2 + Q^2)} = \sqrt{(16 + 27)} = 6.557$
$\tan \theta = Q/P = (3\sqrt{3}.)/4 = 1\cdot299$ and $\theta = 52° 25'$
∴. Resultant is 6.56 N at 52° 25' with 5 N force.

Triangle of Forces

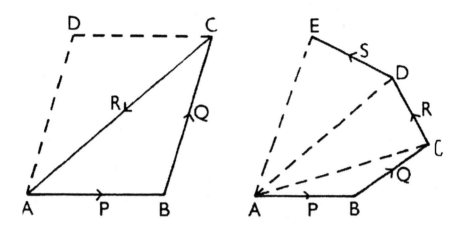

Suppose three concurrent forces P, Q and R can be represented by the sides of the triangle ABC taken in order, i.e. **the arrows go round the triangle in the same sense.** (Fig. 1). Completing the parallelogram $ABCD$, Q is also

represented by AD and hence AC represents the resultant of P and Q. It follows that R is equal in magnitude and opposite in direction to the resultant of P and Q and acts through the same point. Hence the forces P, Q and R are in equilibrium. Thus we have the theorem known as the **Triangle of Forces: If three forces acting at a point are in equilibrium they can be represented in magnitude and direction by the sides of a triangle taken in order.**

If forces P, Q, R, S, ... act on a particle their resultant can be found by extending the triangle of forces to a **Polygon of Forces.** Draw AB, BC, DE, ... to scale to represent P, Q, R, S, ... in magnitude and direction.

AD represents the resultant of P, Q, R; AE represents the resultant of P, Q, R, S, etc. (Fig. 2.) If E **coincides** with A the polygon is **closed** and the **resultant** of the forces P, Q, R, S is **zero**, i.e. **they are in equilibrium.** This holds only if the forces act at the same point.

Thus problems involving forces acting on a particle can be solved by **graphical methods** based on the Triangle and Polygon of Forces. A semi-graphical method can also be used by calculating from a roughly drawn force diagram, but with more than three forces it is usually **easier** to draw an **accurate** diagram and find the answers by **measurement.**

Example. A mass of 5 kg is suspended by two strings inclined at 45° and 30° to the horizontal.

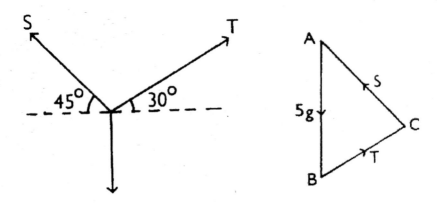

Let S, T be the tensions in the strings in N.

Draw the triangle of forces with AB, BC, CA representing the weight of the mass, 5 g N, T, S respectively. $\angle B = 60°$, $\angle A = 45°$, so $\angle C = 75°$.

Sine Rule: $\dfrac{S}{\sin 60°} = \dfrac{T}{\sin 45°} = \dfrac{5g}{\sin 75°}$

$\therefore S = \dfrac{5g \sin 60°}{\sin 75°} = 43.93$ and $T = \dfrac{5g \sin 45°}{\sin 75°} = 35.88$

where g has been taken as 9.8 m/s^2
$\therefore$ tensions in the strings are 43.9 N and 35.9 N.

LAMI'S THEOREM
If three forces P, Q, R act on a particle and α, β, γ are the angles between Q and R, R and P, P and Q respectively, and the particle is in equilibrium

$$\frac{P}{\sin \alpha} = \frac{Q}{\sin \beta} = \frac{R}{\sin \gamma}$$

In the example above, the angles between the forces are 120°, 135° and 105°. Using Lami's Theorem and $\sin \theta = \sin (180° - \theta)$ leads to

$$\frac{S}{\sin 60°} = \frac{T}{\sin 45°} = \frac{5g}{\sin 75°} \text{ as before}$$

Parallel Forces

The **resultant** R of **two like parallel forces** P and Q is a force **P +Q** in the **same** direction. If P $(>Q)$ and Q are **unlike,** R is equal to **P —Q** and acts in the direction of the **greater** force P. The point at which R acts is found by using the theorem: **The moment of the resultant about any point equals the algebraic sum of the moments of the forces about the same point.**

Let A, B be any two points on the lines of action of P, Q and let the line of action of R cut AB at C. The moment of P about A is zero so the moment of R about A = the moment of Q about A, i.e. the moments have the same magnitude and sense.

If P and Q are **like** forces then Q and R are like forces and Q and R must be on the **same** side of A, i.e. **C is in AB.** If P and Q are **unlike** then Q and R are unlike and Q and R must be on **opposite** sides of A, i.e. **C is in BA produced.** If $Q > P$, R is in the direction of Q and C is in AB produced.

The resultant of any number of parallel forces acting on a body is found by finding resultants of two forces at a time. The final resultant is the **algebraic sum of the forces** and **acts** at a point called the **centre of the parallel forces.**

Example: Consider three equal like parallel forces P acting at the vertices of a triangle ABC.

The resultant of P at B and P at C is 2P at D, the midpoint of BC; the resultant of P at A and 2P at D is *3P* at G on the median AD. Finding the resultants in different order gives G on each of the other medians; hence G is the intersection of the medians.

COUPLES
A pair of equal unlike parallel forces with different lines of action has no single force equivalent and is called a **couple.** The **moment of a couple** is the **same** about any point in the plane of the forces and is equal to the **product of one of the forces and the perpendicular distance between the lines of action of the forces.** In any resolution of forces the forces of the couple will balance and can be neglected, but the **moment** of the couple **cannot** be neglected.

Center of Gravity

A body can be considered to be made up of particles whose weights form a set of parallel forces of which the resultant is the **total weight of the body acting vertically downward** through the center of the forces called, in this case, the **center of gravity** of the body. The center of gravity of a **uniform rod** is at its **midpoint;** of a **uniform rectangular lamina** is at the **intersection of the diagonals;** of a **circular disc** is at its **center;** of a **uniform triangular lamina** is at the **intersection of the medians.**

The center of gravity of a **set of weights acting at points in a plane** is found as follows. Suppose the weights are w_1, w_2, w_3 at the points (x_1, y_1), (x_2, y_2), (x_3, y_3) referred to rectangular axes OX, OY. The resultant is $w_1 + w_2 + w_3$ acting at G $5/$) and the moment of the resultant about the axes is equal to the sum of the moments of the weights about the axes.

Moments about $0Y$: $(w_1 + w_2 + w_3)x = w_i x_i + w_2x_2 + w_3x_3$
Moments about ON: $(w_1 + w_2 + w_3) y = w_1y_1 + w_2y_2 + w_3y_3$
Hence (x, y) , the center of gravity.
The same method is used to find the center of gravity of a compound body, the weights of the constituent bodies being taken at their centers of gravity.

CENTER OF GRAVITY OF REMAINDER

Let W, W_1, W_2 be the weights of the original body, the part removed and the remainder, and G, G_1, G_2 the corresponding centers of gravity. W at G is the resultant of W_1 at G_1 and W_2 at G_2. The moment of W about G is zero, so the moments of W_1 and W_2 about G are equal but opposite in sense, i.e. **G is between G_1 and G_2 and $W_2.GG_2 = W_1GG_1$**

Example: A circular hole of radius 2 cm is drilled 5 cm from the center of a circular plate of radius 8 cm. If w is the weight per unit area,
$W = 64\pi w$, $W_1 = 4\pi w$, $W_2 = 60\pi w$, and $GG_1 = 5$ Moments of W_2, W_1 about G are equal.
Hence $60\pi w$ $GG_2 = 4\pi w$ x 5 and $GG_2 = 0.33$cm $= 33$mm

A **suspended** body will hang with its **centre of gravity vertically below the point of suspension.**

Equilibrium

A body acted on by forces is in equilibrium if :

I.   **The algebraic sum of the resolved parts of the forces in any two directions is zero.**

II.   **The algebraic sum of the moments of the forces about any point is zero.**

Alternatively, we can **resolve in one direction and take moments about each of two points** or we can **take moments about each of three non-collinear points.** In general, I and II are the **best** conditions to use. In any case we obtain **three** equations which are sufficient to determine **three** unknowns.

For a **particle** we resolve to get **two** equations only.

If a body is in equilibrium under the action of **three** forces the forces are **coplanar** and must be **either parallel or concurrent.** If the forces are **concurrent** the equilibrium problem can be treated as for a particle by two resolutions or by the triangle of forces.

In all problems of the equilibrium of a body a clear figure should be drawn and **all the forces marked in.** Do not overlook the reaction between bodies in contact.

FRICTION

When bodies are in contact the reaction between them depends on the kind of surfaces. For a body on a **smooth** surface the **reaction is normal to the surface.**

If the contact is **rough,** the **total reaction (S)** has two components; one **normal** to the surface and called the **normal reaction (R)** and the other **along** the surface **opposing the tendency to move** and called the **friction force (F). In** equilibrium the friction force is **just enough** to prevent motion.

If a body is **just about to slide** the equilibrium is said to be **limiting** and the friction force is then a **maximum** related to the normal reaction by F = **µR,** where μ **is a constant called the coefficient of friction.** In equilibrium problems involving friction we therefore have two additional unknowns connected by the relation $F = \mu R$.

It is sometimes convenient to use the total reaction; when the equilibrium is limiting the total reaction makes an angle λ with the normal where **tanλ =μ,** and λ is called the **angle of friction.**

Example: A uniform ladder 10 m long and weighing 24 kg rests with one end on rough horizontal ground and the other against a smooth vertical wall. When about to slip the ladder makes an angle of 30° with the wall.

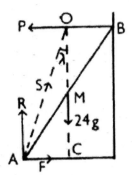
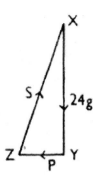

The foot of the ladder tends to slip away from the wall so the friction force F acts towards the wall. Let R be the normal reaction at the ground and S the total reaction at an angle λ to the vertical. The equilibrium is limiting so the coefficient of friction $\mu = \tan \lambda$. The wall is smooth so the reaction at the wall, P, is perpendicular to the wall, i.e. horizontal.

The forces on the ladder are therefore as marked. Since the ladder is in equilibrium,

Resolving horizontally: $F = P$

Resolving vertically: $R = 24g$

Moments about A: P x 10 cos 30° = 24g x 5 sin 30°

Hence $P = 4g\sqrt{3}$ N, $F = 4g$ N$\sqrt{3}$, $R = 24g$ N and can be further evaluated by taking $g = 9.8$ m/s^2.

$S = (F^2 + R^2) = g (48 + 576) = g,/624 = 245$ N.

$\mu = \tan = F/R = 4g/3/24g = \sqrt{3/6}= 1$ x 732 ÷ 6 = 0.2887

∴ $\mu = \dfrac{\sqrt{3}}{6}$ and $\lambda = $ **16° 6'**

Otherwise, we can treat this as a three force problem by using the total reaction S instead of F and R. For equilibrium S must go through 0, the intersection of P with the line of action of the weight. The triangle of forces XYZ is similar to triangle OCA, so $\angle X = 2$.

∴ $P = 24g \tan \lambda$ and $24g = S \cos \lambda$.

Also AC OC $\tan\lambda$, and $AC = \dfrac{1}{2}OC \tan 30°$ since $AM = MB$.

Hence λ, μ, P and R can be calculated.

Uniformly Accelerated Motion

Denoting the distance by s and the time by t then $\frac{ds}{dt}$ is the velocity v and $\frac{d^2s}{dt^2}$ is the acceleration a.

A negative acceleration is **retardation**.

In **SI units**, the unit of velocity is the **meter per second, i.e. m/s**, and the unit of acceleration is the **meter per second per second, i.e. m/s²**.

If velocity is plotted against time, the **gradient** of the curve obtained is $\frac{dv}{dt}$, i.e. the **acceleration**.

If the acceleration is **constant** the graph is a **straight line.** If the velocity is constant the graph is a straight line **parallel to the time axis.**

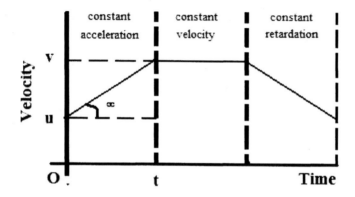

$\frac{ds}{dt} = v$, so $s = \int v \, . \, dt$. The integral $\int_{t_1}^{t_2} v \, . \, dt$ is the area under the curve. Hence the **area under the velocity-time curve between the ordinates t = t₁ and t = t₂ is numerically the same as the distance gone in the time t₂ —t₁.**

Suppose that when $t = 0$, $s = 0$, $v = u$ (the initial velocity) and that the acceleration a is constant.

From the velocity-time graph for constant acceleration

the gradient $= a = \tan \alpha = \frac{v-u}{t}$

Hence   **v = u +at**

Area under the graph = area of trapezium $= \frac{1}{2}(u + v).t$ **(1)**

Hence $s = \frac{u+v}{2}.t$ **(2)**

Eliminate *t* **between (1) and (2):** $\frac{v-u}{a} = \frac{2s}{v+u}$

Hence $v^2 - u^2 = 2as$ **(3)**

Eliminate *v* **between (1) and (2):** $s = \frac{1}{2}(2u + at).t$

Hence $s = ut + \frac{1}{2}at^2$ **(4)**

(1), (2), (3), (4) are the equations for uniformly accelerated motion and each involves **four** of the **five** quantities **s, u, v, t, a**. Hence, given any three the remaining two can be found.

Example: A particle moves in a straight line with constant acceleration. In the first 2 seconds it travels 28 m and in the next 2 seconds it travels 24 m. Find the initial velocity and the retardation.

With the usual notation, $s = 28$ when $t = 2$, and $s = 52$ when $t = 4$.

Using $s = ut + \frac{1}{2}at^2$,

$28 = 2u + 2a$

$52 = 4u + 8a$, i.e. $13 = u + 2a$

$\therefore 2a = 28 - 2u = 13 - u$, i.e. $u = 15$.

Substituting for u: $a = (13 - 15)/2 = -1$.

$\therefore$ Initial velocity = **15** m/s and retardation = 1 m/s^2.

MOTION UNDER GRAVITY

A body allowed to fall freely towards the earth **falls with a constant acceleration (g)** usually taken as 9.80 or **9.81 m/s**^{2}. The equations of motion are therefore those of uniformly accelerated motion.

Example: A ball is thrown vertically upward with a velocity of 14 m/s. Find the greatest height reached and the time before the ball returns to hand. For how long is the ball more than 7 m above the level from which it was thrown ? (Take $g = 9\cdot8$ m/s^2)

Taking the upward direction as positive, $a = -9\cdot8$. $v = 0$ at the highest point and $u = 14$.

$v^2 - u^2 = 2as$ gives $0^2 - 14^2 = -19.6 \times s$

and s $= 196 \div 19\cdot6 = 10$

        $\therefore$ greatest height reached = 10 m

MOTION UNDER GRAVITY continued.

$s = 0$ when the ball returns to hand and using

$s = ut + \frac{1}{2}at^2 : 0 = 14t-4.9t^2 = 7t(2 - 0.7 \text{ x } t)$

$\therefore t = 0$ (projection) and $20 \div 7 = 2.857$

$\therefore$ time to return to hand = 2.86 s.

The ball is at height 7 m going up and coming down.

$s = ut + \frac{1}{2}at^2: 7 = 14t - 4.9 .t^2$ or $7t^2 -20t+10 = 0$

Solving this quadratic equation by use of the formula, the roots are 0.647 and 2.211 and the ball is at height 7 m after 0.647 s and again after 2.211 s.

$\therefore$ Ball is above 7 m for 1.56 s.

PROJECTILES

If a body is projected with velocity V at a n angle α to the horizontal, the **horizontal component of the velocity,** $V \cos \alpha$, is **constant throughout** the flight, but the **vertical component,** $V \sin \alpha$, **changes** owing to the acceleration due to gravity. The horizontal motion is constant velocity motion, the vertical motion is motion under gravity, and the two motions are considered **independently.**

Example: A particle is projected with velocity 24.5 m/s at an angle of 30° with the horizontal. $(g = 9.8$ m/s$^2)$ Horizontal and vertical components of velocity of projection are $(24.5 \text{ x } \cos30°, 24.5 \text{ x } \sin30°)$ i.e. $12.25.\sqrt{3}$ m/s and 12 .25 m/s respectively.

Let x, y m be the horizontal and vertical distances and v m/s the upward vertical velocity at time t s.

For the horizontal motion: $x = 12.25 \text{ x } \sqrt{3} \text{ x } t$ (1)

for the vertical motion: $u = 12 .25, a = -9.8$

$v = u + at$ gives $v = 12.25 - 9.8 \text{ x } t$ (2)

$s = ut + \frac{1}{2}at^2$ gives $y = 12.25 \text{ x } t-4-9 \text{ x } t^2$ (3)

Particle strikes ground when $y = t(12.25 -4.9 \text{ x } t) = 0$

i.e. $t = 0$ (projection) or $12.25 \div 4·9 = 2 \text{ x } 5$.

$\therefore$ time of flight = 2.5 s.

Putting $t = 2.5$ in (1), horizontal range = 53.0 m.

At highest point $v = 0$ and (2) gives $0 = 12 .25 - 9-8 .t$.

Time to highest point $\qquad = 1.25$ s

$\qquad = \frac{1}{2} \text{ x time of flight.}$

From (3), greatest height $= 12.25 \text{ x } 1.25- 4.9 \text{ x } (1.25)^2$

$\qquad = 15.31 -7.656$, by logs.

$\therefore$ Greatest height $\qquad = 7.65$ m.

Laws of Motion

Newton's Laws of Motion state that:

I. Every body continues in its state of rest or of uniform motion in a straight line unless it is compelled to change that state by external forces.

II. The change of motion is proportional to the applied force and takes place in the direction of the force.

III. To each action there is an equal and opposite reaction.

I gives, in effect, a **definition of force** and from **II** we deduce that the **resultant force** acting (F) is **proportional** to the **acceleration** produced (a), i.e. $F = ka$. Taking $k = m$, the measure of the mass, then unit force gives unit mass unit acceleration. In SI units, **unit force is a Newton (N) which when applied to a mass of 1 kg gives it an acceleration of 1 m/s².**
It follows that the **weight** of a mass m kg is **mg N**.
The relation **F = ma** is a fundamental equation in dynamics.

Example: Consider a mass of 6 kg on a smooth horizontal table connected by a light inextensible string to a mass of 4 kg hanging freely over the table edge.

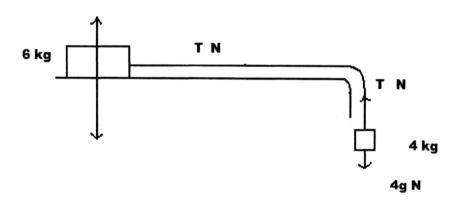

When the system is released let its acceleration be a m/s² and the tension in the string be T N. The net downward force on the hanging mass = $(4g - T)$ N and the horizontal force on the mass on the table = T N.
Writing the equation of motion, $F = ma$, for each mass: $4g - T = 4a$ and $T = 6a$, and a, T are easily found.

Work and Power

The **work** done by a force is the **product of the force and the distance moved by its point of application in the direction of the force** and is measured in **joules,** symbol **J,** when the force is in newtons and the distance in meters. The work done by a horizontal force of **10** N in moving a body. 5 m horizontally is 10 x 5 = 50 J if the body is pushed 5 m up a plane inclined at an angle α to the horizontal and the horizontal distance moved is 5 cos α m then the work done by the horizontal force of **10** N is 10 x 5 cos α = 50 cos α J.

Power is the rate at which work is done. The unit of power is the **watt,** symbol W, and is equal to **1 joule per second,** i.e. **1 J/s.**

Example: A train of 200 tons is travelling at 24 km/h along a level track against resistances of 72 N per ton.
Resistances = 72 x 200 = 14 400 N.
Speed = 24 x 1 000 ÷ 3, 600 = 20/3 m/s.

Work done per sec. against resistances = 14 400 x $\frac{20}{3}$ J

i.e. power required = 96 000 W = 96 kW.

N.B. 1 kW = 1 kilowatt = 1000 W.
If the train is moving up a slope of 1 in 160 the resistance is increased by the component of the weight down the plane. 1 ton = 1 000 kg so the weight of the train = 200,000g N and its component down plane is
$\frac{1}{160}$ x 200,000 x 9.8 = 12,250 N
Total resistance = 26,650 N
i.e. power required = 26,650 x $\frac{20}{3}$ = 177666 W
= 178 kW

NUMERICAL VALUES

When a quantity is expressed by a numerical value and a unit it is generally preferable to use units so that the numerical value lies between 0.1 and 1000. To obtain the numerical value in this range the unit chosen is a decimal multiple or sub-multiple of the SI unit formed by **means of a prefix.**

Energy

The energy of a body is its **capacity for doing work. Kinetic Energy (K.E.)** is the energy it possesses by virtue of its **motion** and is equal to $\frac{1}{2}mv^2$ J where m is its mass in kg and v its velocity in m/s.

If a force F gives a mass m an acceleration a which increases its velocity from u to v in a distance s, we have

$$F = ma \text{ and } v^2 - u^2 \; 2as.$$

Eliminating a between these equations gives

$$Fs = \frac{1}{2}mv^2 - \frac{1}{2}mu^2$$

i.e. **the work done by the applied force is equal to the change in the kinetic energy.**

Example: A 0.03 kg bullet moving at 500 m/s penetrates a fixed block of wood to a depth of 200 mm.

Let $F\,N$ be the average resistance of the wood.

Work done by resistance = loss of K.E. of the bullet

$$\therefore F \; \text{x}0.2 = \frac{1}{2} \text{x}\,0.03 \; \text{x}500^2 = \frac{1}{2} \text{ x 3 x 2500}$$
$$\therefore F = 7\,500 \div 0.4 = 18\,750 \text{ N} = 18.75 \text{ kN}.$$
$$\therefore \text{average resistance of wood} = 18\cdot75 \text{ kN}.$$

The **Potential Energy (P.E.)** of a body is the work that the forces acting on it would do if it moved from its **actual position** to **some standard position**. It is the energy possessed by virtue of **position.** For a. body of mass **m kg** at a height **h m** above ground level the potential energy **is mgh J.**

Example: A mass m kg falls from a height h m. Initially the P.E. $= mgh$ J and it has no K.E.

When it has fallen a distance s m the P.E. $= mg(h - s)$ The velocity at this point is given by $v^2 = 2gs$

$\therefore$ at this point, K.E. $= \frac{1}{2}mv^2 = \frac{1}{2}m \text{ x } 2gs = mgs$

and P.E. + K.E. $= mg(h - s) + mgs = mgh.$

i.e. **the sum of the potential energy and the kinetic energy is constant.**

This is an example of the principle known as the **Conservation of Energy.** The Conservation of Energy principle must **not** be used when there are frictional forces, impacts, etc., because some of the kinetic energy then lost is transformed into other forms of energy, mainly heat.

Momentum and Impulse

If a force F gives a mass m an acceleration a which increases its velocity from u to v in time t, we have

$$F = ma \text{ and } v = u + at.$$

Eliminating a between these equations and re-arranging gives $Ft = mv - mu$. The quantity **mv** is the **momentum of the body** and the quantity **Ft is** the **impulse of the force**, both measured in **kg m/s**. Since velocity is a vector both **momentum and impulse are vectors**. The equation above is stated. **Change in momentum = impulse of the applied force.**

Example: A mass of 4 kg at rest is struck and moves off with a velocity of 3 m/s. If the blow lasts for 0.02 s, calculate the average force.
Let the average force be F N.
Impulse of applied force = change of momentum
$\therefore F \times 0.02 = 4 \times 3$ and $F = 600$ N.

If two bodies moving in the same straight line collide there will be a short period of time in which there will be an equal and opposite reaction between the bodies. Thus the momentum **lost** by one is equal to the momentum **gained** by the other. If m_1, m_2 are the masses, u_1, u_2 and v_1, v_2 the velocities before and after the collision respectively,

$$m_1u_1 - m_1v_1 = m_2v_2 - m_2u_2$$
$$\therefore m_1u_1 + m_2u_2 = m_1v_1 + m_2v2$$

Thus **the momentum before impact = the momentum after impact, i.e.- the linear momentum is conserved.**

Example: A 4 kg mass moving with velocity 4 m/s collides with a 6 kg mass moving in the opposite direction with velocity 1 m/s and they move off together.
Take direction of the first mass as the positive direction and V m/s as the common velocity after collision.
Momentum after collision = momentum before collision
$\therefore (4 + 6)V = 4.4 + 6.(-1) = 10$; hence $V = 1$ m/s.
K.E. before collision $= \frac{1}{2} . 4 . 4^2 + \frac{1}{2} . 6 . 1^2 = 35J$
K.E. after collision $= \frac{1}{2} . 10 . 1^2 = 5J$
$\therefore$ **K.E. is lost during the collision.**